Lecture Notes in Networks and Systems

Volume 112

The series "Lecture Notes in Networks and Systems" publishes the latest developments in Networks and Systems—quickly, informally and with high quality. Original research reported in proceedings and post-proceedings represents the core of LNNS.

Volumes published in LNNS embrace all aspects and subfields of, as well as new challenges in, Networks and Systems.

The series contains proceedings and edited volumes in systems and networks, spanning the areas of Cyber-Physical Systems, Autonomous Systems, Sensor Networks, Control Systems, Energy Systems, Automotive Systems, Biological Systems, Vehicular Networking and Connected Vehicles, Aerospace Systems, Automation, Manufacturing, Smart Grids, Nonlinear Systems, Power Systems, Robotics, Social Systems, Economic Systems and other. Of particular value to both the contributors and the readership are the short publication timeframe and the world-wide distribution and exposure which enable both a wide and rapid dissemination of research output.

The series covers the theory, applications, and perspectives on the state of the art and future developments relevant to systems and networks, decision making, control, complex processes and related areas, as embedded in the fields of interdisciplinary and applied sciences, engineering, computer science, physics, economics, social, and life sciences, as well as the paradigms and methodologies behind them.

** Indexing: The books of this series are submitted to ISI Proceedings, SCOPUS, Google Scholar and Springerlink **

More information about this series at http://www.springer.com/series/15179

Alexánder Martínez · Héctor A. Moreno ·
Isela G. Carrera · Alexandre Campos ·
José Baca

Editors

Advances in Automation and Robotics Research

Proceedings of the 2nd Latin American Congress on Automation and Robotics, Cali, Colombia 2019

 Springer

Editors
Alexánder Martínez
Pontificia Universidad Javeriana de Cali
Cali, Valle del Cauca, Colombia

Héctor A. Moreno
Universidad Autónoma de Coahuila
Monclova, Coahuila, Mexico

Isela G. Carrera
Universidad Autónoma de Coahuila
Monclova, Coahuila, Mexico

Alexandre Campos
Universidade do Estado de Santa Catarina
Balneário Camboriú, Santa Catarina, Brazil

José Baca
Texas A&M University – Corpus Christi
Corpus Christi, TX, USA

ISSN 2367-3370   ISSN 2367-3389   (electronic)
Lecture Notes in Networks and Systems
ISBN 978-3-030-40308-9   ISBN 978-3-030-40309-6   (eBook)
https://doi.org/10.1007/978-3-030-40309-6

This Springer imprint is published by the registered company Springer Nature Switzerland AG
The registered company address is: Gewerbestrasse 11, 6330 Cham, Switzerland

Preface

This book contains a selection of the contributions reported at the 2nd Latin American Congress on Automation and Robotics, LACAR 2019. This book presents recent advances in the analysis, design, control and development of autonomous and robotic systems, which are intended for a variety of purposes such as manufacturing, agriculture, health care, rehabilitation, locomotion, computer vision and education.

LACAR 2019 was held at the Pontificia Universidad Javeriana, in Cali, Colombia, from October 30 to November 1, 2019. This conference promotes an open forum where researches, scientists and engineers come together to present the results of their research. LACAR conference aims to generate activities in research, development and application on automation and robotics by exchanging knowledge, experiences and the synergy of research groups from different places from Latin America.

We are grateful to the authors of the articles for their valuable contributions, and to the reviewers who donated their time to revise them. We want to acknowledge and thank the members of the organizing committee for all their hard work and dedication, which made this possible. We would also like to extend our sincere gratitude to the editorial staff of Springer for supporting this project.

Contents

Industrial Robots Migration Towards Industry 4.0 Components

Juan David Contreras[✉]

Pontificia Universidad Javeriana, Cl 18-118-250, Cali, Colombia
juandavid.contreras@javerianacali.edu.co

Abstract. The use of industrial robots rises as a key element for the fourth industrial revolution also known as industry 4.0. Nevertheless, a common characteristic of current industry 4.0 applications and robotic systems is the lack of a proper standardization framework which is one of the pillars of said revolution. In response, this paper proposes a procedure for turning industrial robots into Industry 4.0 components according to the guidelines of the Reference Architecture Model for Industry 4.0 (RAMI 4.0). The defined procedure is applied for two different industrial robots leading to robotic systems that can be integrated into smart factories of industry 4.0 with new levels of interoperability. This allows to offer new value opportunities based on data analysis.

1 Introduction

The fourth industrial revolution or Industry 4.0 is an initiative from the European Union to increase the value added of the industry by applying information and communication technologies (ICT) throughout the entire value chain [6]. Although Industry 4.0 has received different names from one country to another, it is commonly characterized by a migration process from the traditional industry to a network of smart factories. In said network, the components of the value chain are intertwined in the digital world leading to Cyber-Physical Systems (CPS) which communicate with each other through the Internet of Things (IoT) [11]. This new approach for the organization and implementation of manufacturing systems offers new capabilities such as decentralized decision-making, mass customization, horizontal and vertical integration and end-to-end engineering [12].

The concept of Industry 4.0 has been mostly associated with solutions and technologies that are expected to be implemented, such as additive manufacturing, Collaborative Robots (Cobots), Big Data and cloud computing, among others. This perspective on Industry 4.0 although correct is incomplete. Not every implementation of these technologies in industrial scenarios can be considered as a proper example of Industry 4.0. To avoid such misconceptions, it is necessary to follow a set of implementation rules for the communication and virtual representation of robotics systems defined by the Plattform Industrie 4.0.

Traditionally, the use of robots in factories has become the first alternative chosen to handle tasks that may be dangerous for mankind, to achieve faster

A. Martínez et al. (Eds.): LACAR 2019, LNNS 112, pp. 1–12, 2020.
https://doi.org/10.1007/978-3-030-40309-6_1

and more accurate production processes, and to reduce the overall cost of products. Moreover, given the increasing competitiveness in today's globalized markets, manufacturers need to improve their decision-making processes by using tools such as artificial intelligence and Big Data. In addition, more efficient, autonomous, and customizable processes are currently required in terms of new levels of automation and human-machine integration [10]. Consequently, in the context of Industry 4.0, modern progress in information technologies will affect the use and design of robots within the industry in order to respond to the new requirements of a global economy [4].

Based on the previous statement, the use of industrial robotic manipulators and automatic guide vehicles (AGV) becomes a key element for the fourth industrial revolution [10]. Several publications have been presented in recent years (2014–2019), thus showing the effect of Industry 4.0 in the field of robotics. For example, Wurtz and Golz [16] propose an upgrade to traditional Matlab tools for the integration of TCP/IP protocols during the maneuver of wireless and remote controlled robots via the internet. In this sense, Bohuslava et al. [7] designed a communication strategy via TCP/IP and socket technology for the control of robotic cells. The work presented by Mario et al. [13] discusses a CPS-based approach for the modelling and co-simulation of AGVs in Smart Factories. On a different note, Cobots can be used to support human workers in Industry 4.0 manufacturing environments [8] while the Cloud Robotics (CR) paradigm allows robots to benefit from the computational, storage and communication resources provided for the Industry 4.0 ICT infrastructure [2]. Nevertheless, a common aspect of these contributions is the lack of a standardization framework, which is one of the pillars of Industry 4.0 [15]. This implies that not all applications can be accepted as proper solutions within the context of Industry 4.0 due to the importance of standardization regarding its purpose.

Considering the opportunities of Industry 4.0 in the field of robotics, the objective of this paper is to present the requirements that an industrial robot must fulfil in order to be considered an Industry 4.0 Component (I4C). This includes the definition of both the standardization framework and the ICT hardware requirements for the communication and virtual representation of industrial robots, according to the rules and guidelines established by the Plattform Industrie 4.0. As an example, the migration of an industrial robot towards Industry 4.0 is also showcased.

2   Industry 4.0 Components

2.1   Reference Architecture Model for Industry 4.0

Since 2015, technical reports have been published by Plattform Industrie 4.0, beginning with the Reference Architecture Model for Industry 4.0 - RAMI 4.0, which was presented as a guideline that "permits step-by-step migration from the world of today to that of Industry 4.0" [1]. RAMI 4.0 is a three-axes model, where each axis represents a domain or point of view in the context of Industry 4.0. The vertical axis represents the information and communication technology

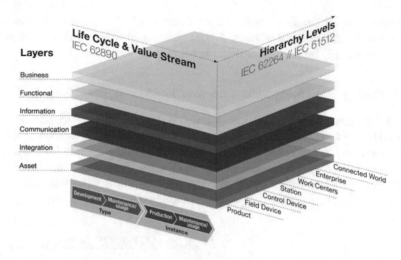

Fig. 1. Reference Architecture Model for Industry 4.0.

(ICT) domain divided into several layers. In the left-most horizontal axis, the product life cycle management is represented according to IEC 62890 standard. Finally, the right-most horizontal axis represents a functional hierarchy based on the IEC 62264 standard with the addition of the elements from 'Connected World', 'Field Device' and 'Product', see Fig. 1.

Industry 4.0 Components (I4C) are defined in RAMI 4.0 as a combination between a thing (e.g. Machine, field device, factory, software) and an Administration Shell (14), see Fig. 2. In the vertical axis of RAMI 4.0, the Administration Shell will include functional, information and communication layers.

2.2 Details of the Asset Administration Shell

The Administration Shell is described as the virtual representation of things, assets and objects including software, documents, people, etc. When the administration shell corresponds to the virtual representation of a physical asset, it is known as an Asset Administration Shell (AAS). According to the previous statement, the correct implementation of an Industry 4.0-compliant robot depends on the correct implementation of an AAS and an Industry 4.0-compliant communication system (Fig. 1). In other words, a set of standards for the implementation of information models (i.e. structured data and services) must be fulfilled.

The basic structure of the AAS is presented in Fig. 3 where the AAS is divided into: a Header, where the identifications of both the asset and the AAS are stored, and a Body in which the sub-models, virtually representing the asset, are stored (15). The main parts of an AAS are the represented asset and the sub-models. Figure 4 shows a list of several standardized sub-models that can be used in the AAS depending on the application. For instance, said models may contain a description regarding the safety or security defined to aid operators, and present a sub-model for condition monitoring which assesses the state of the asset and prevents possible failures. A detailed specification of the structure and implementation of the Administration Shell can be found in: [3].

2.3 Industry 4.0 Compliant Communication

RAMI 4.0 defines the IEC 62541 standard (an open-standard UA protocol managed by the OPC Foundation) as the approach used for the implementation of the communication layer. OPC UA is also targeted as a format for information models and communications regarding production operations for AAS implementation [3]. This implies that Industry 4.0-compliant communications must be implemented throughout the productive system using OPC UA.

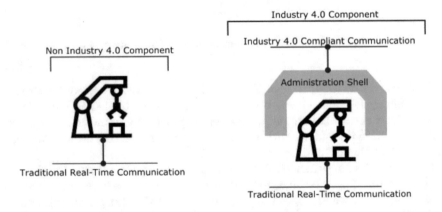

Fig. 2. Comparison between Industry 4.0 components and non-components.

OPC UA works as a platform-independent technology for industrial communication over a TCP protocol. Additionally, OPC UA can run on both embedded systems and enterprise systems. The potential of OPC UA consists in the combination of transportation mechanisms and data modeling to provide server-client communication where the information on the server side is published using Type-Instance modelling rules and base modelling constructs thus forming a structured information model. This approach enables an OPC UA client to access the smallest pieces of data without the need to understand the entire model, especially in complex systems (18).

Type Instance data modelling of OPC UA has helped other organizations to build their models on top of the OPC information model, allowing standards such as PLCopen, FDI, FDL and AutomationML to be mapped into OPC UA. In this sense, standardized communication is possible between users and information providers (19).

According to [14], an AAS can be implemented by using OPC UA by following four migration steps:

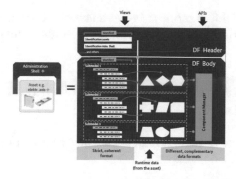

Fig. 3. Structure of the Asset Administration Shell.

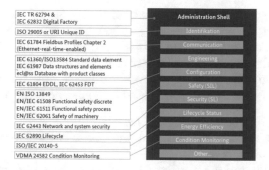

Fig. 4. Standardized sub-models used in AAS.

- **Migration Step 1:** Information Access. OPC UA is used to provide simple communication between devices by allowing information access. The information model for OPC Device Integration (DI) exposes variables provided by machines and plants. Said variables can be found and manually subscribed to.
- **Migration Step 2:** Companion Specification. In order to increase interoperability and enable plug & produce applications, a standardized information model is used to describe some of the general characteristics and functionalities of devices and the topologies in which they might operate. This model is referred to as Companion Specification and involves the definition of branch-specific devices and their applications, e.g. field devices, PLCs or machines.
- **Migration Step 3:** Extended Information Model. In addition to several companion specifications, an unstandardized information model can be added. The extended information model includes information and features that the developer wishes to offer explicitly and pertain to experience and proprietary knowledge.
- **Migration Step 4:** Asset Administration Shell. The data elements and services on the AAS should be standardized using properties and service specifications. On top of additional companion specifications, this also involves

standardized individual terms that can be used in information models. Global identifiers (e.g., IRDI) should be used for interoperability between I4.0 components.

3   Migration of Industrial Robots Towards Industry 4.0 Components

On order to retrofit or design an industrial robot that can be considered an I4C, the following procedure is proposed.

3.1   Design and Implement the ICT Architecture

Depending on the computational capacity of the robot, a decision must be made on where the AAS will be implemented. Although modern robots have the ability to store information and run OPC UA clients and servers within their built-in controllers, older robots must be retrofitted in order to guarantee sufficient computational capabilities [9]. In general, the robot must include the following capabilities, given that each one is related to a specific layer of the RAMI 4.0:

- IP communication (communication layer).
- Sufficient data storage in relational and non-relational databases, XML or JSON (Information layer).
- Data acquisition and control through physical ports such as GPIO ports or serial communication, as well as other types of digital or analog inputs and outputs (integration layer).
- Information processing through software functions for the implementation of high-level services (functional layer).

The design of the ICT architecture consists on adjusting the previously mentioned capabilities to the robotic system. This architecture can be either physical (e.g., single-board computer) or virtual (e.g., cloud computing). Depending on the system requirements, field devices and ICT hardware should be implemented.

3.2   Configure Communication with OPC UA

OPC UA servers and clients should be deployed by using the implemented ICT architecture. Initially, a generic information model such as the OPC UA Device Integration (OPC DI) is chosen to publish identification data such as the robot manufacturer, its model, and serial number. Furthermore, application-specific parameters and methods can be loaded to the server within this model, thus allowing external I4C to browse, read, write and call methods as well as subscribe to variables and events.

3.3   Populate the Information Model with Companion Specifications

Companion specifications are standardized information models for particular applications. In the case of robotic applications, a first companion specification was published in 2018. This information model provides access to asset management and data from motion device systems for condition monitoring. Based on the provided data, the following use cases are supported:

1. Supervision: The robot can be supervised and monitored. During the production phase, the model provides data regarding the operational and safety states as well as process-related data.
2. Condition monitoring: The data is used to determine the condition of the robot during operation. This enables the identification of significant changes which may be indicative of developing faults.
3. Asset management: This includes information on the main electrical and mechanical parts such as part number, brand name and serial number. These data may indicate whether maintenance is required, given that the technical expert knows in advance which parts need to be replaced and prepare accordingly.

An example of the communication structure based on companion specification is presented in Fig. 5. In this application, real-time data and parameters are mapped from the robot to OPC UA. Then, the structured data is read from a database to be used by a data analysis application.

The companion specification for robotics is shown in Fig. 6, where some elements are detailed as well as the parameters involved in the description of the Motion Device. In this figure, the *MotionDeviceSystem* is shown as the core element of the model which include three folders for *MotionDevices*, *Controllers* and *SafetyStates*, as a example, the *MotionDevices* folder is detailed which include properties such as *Model*, *Axes* and *FlageLoad* among others. The detailed description of all the elements of the model can be found in the document "OPC UA Companion Specification for Robotics (OPC Robotics) – Part 1: Vertical integration".

Furthermore, other companion specifications can be used depending on the applications aimed by the developers (see Fig. 2).

3.4   Additional Information and Services

To complete the implementation of the AAS for a robotic system, other data elements need to be added such as global identifiers, individual standardized properties and service specifications, references, views and unstandardized information for local applications [5].

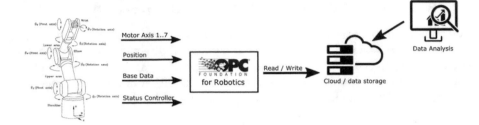

Fig. 5. Communication structure for data analysis using OPC UA.

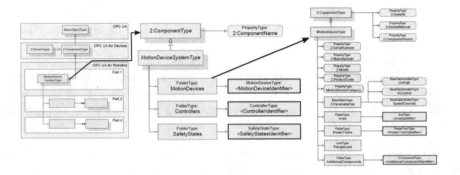

Fig. 6. OPC robotics information model overview.

4   Application Scenario

In order to exemplify the migration process (Sect. 3), two robots are used as application examples. The first robot is a Mitsubishi Melfa RV-M1, which is a model from 1987 used since 1998. This robot can be programmed and controlled only through a RS232 serial interface. It is currently used as a system for material handling in a CNC machine (Fig. 7).

The second robot is the UR3, a 2018 model from Universal Robots in use since 2019. This cobot can be controlled and programmed directly from its teaching pendant or remotely via Modbus TCP or Ethernet/IP. The UR3 robot is currently used in an assembly station (Fig. 8). Both the Melfa RV-M1 and UR3 models are part of the same flexible manufacturing cell at Pontificia Universidad Javeriana in Cali, Colombia.

The Melfa RV-M1 model has several limitations due to its ageing technology. Although real-time data cannot be obtained from motor loads or joint positions, coordinate positions and status codes can still be read. Hence, an external computational system is required to deploy an OPC UA server and manage serial communication through a subprogram. The architecture is shown in Fig. 9, using a Raspberry Pi 3 as the computational system, the historical data including operation time and counting-related parts are stored in a database and the parameters are stored in XML files (information models). The *Serial Commands*

Fig. 7. Mitsubishi Melfa RV-M1

Fig. 8. UR3 robot for the application example.

Manager is a script in charge of mapping the method calls from the OPC UA server to robot commands and copying the data read from the robot controller into the database.

On the other hand, the UR3 has a modern controller which integrates a Mini-ITX PC with a Debian operating system. This enables the OPC UA server to be installed into the robot controller and perform a direct mapping of the robot variables to the information model. Figure 10 shows the proposed ICT architecture to deploy the AAS. The OPC UA server is populated by an internal script called the variable mapper. The server can operate alongside the native communication of the robot by sharing the Ethernet port and the IP, but using a different network port.

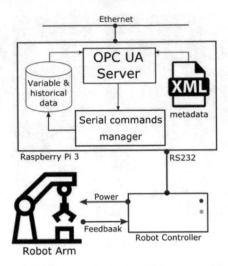

Fig. 9. Proposed ICT architecture for Melfa RV-M1.

In both models, the OPC UA server is implemented using the *freeopcua* Python library. The corresponding parameters of the robots are initially written to the OPC DI as well as the companion specification for robotics using Eclipse XML Editor. Afterwards, the XML files are loaded into the information model of the OPC UA server using the *import_xml* function. Finally, the variables and methods are linked to Python scripts which are respectively in charge of updating the data and activating the robot functions. Comparing the results obtained, it is determined that the RV-M1 robot has a great limitation due to the communication interface between the robot and the OPC UA server since the robot ignores any message receive while it is executing any command, this forces that in the programming of the *Serial Command Manager*, recursive methods are used to know when the robot is available. On the contrary, the UR3 allows to easily map the robot variables from the Modbus registers to the information model on the OPC UA server without interfering with the operation of the robot, but it is required to schedule an agency of services when multiple OPC UA clients requires services that involve movements. In both cases, having access to structured data about service availability, operating time or response times allows high-level applications that integrate OPC UA clients to improve decision making at an operational and logistic level.

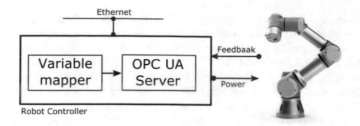

Fig. 10. Proposed ICT architecture for UR3.

5   Conclusions

A procedure for the migration of industrial robots towards Industry 4.0 was proposed based on the formal guidelines and standards provided by Plattform Industrie 4.0. Hence, the developed robotic systems can be integrated into more complex Industry 4.0 systems such as plants, manufacturing cells or modular stations.

The use of standardized information models and communication protocols in the virtualization of industrial robots allows to develop new applications that can harness new levels of interoperability and available data. Modern tools such as machine learning can be used to improve the manufacturing processes and find new value opportunities within robotic applications.

The current setup still has its own limitations, which are the subject of further research. The next step is to develop a framework that reduces the time invested in monotonous and repetitive tasks. For instance, this may involve the one-by-one binding of the database variables with their respective counterparts in the OPC UA information model.

References

1. Adolphs, P., Bedenbender, H., Dirzus, D.: Reference Architecture Model Industrie 4.0 (RAMI4.0). Tech. rep., VDI/VDE Society for Measurement and Automatic Control (2015)
2. Aissam, M., Benbrahim, M., Kabbaj, M.N.: Cloud Robotic: Opening a New Road to the Industry 4.0, pp. 1–20. Springer, Singapore (2019)
3. Barnstedt, E., Bedenbender, H., Billmann, M.: Details of the asset administration shell. Tech. rep., Federal Ministry for Economic Affairs (2018)
4. Bayram, B., İnce, G.: Advances in Robotics in the Era of Industry 4.0, pp. 187–200. Springer, Cham (2018)
5. Bedenbender, H., Bentkus, A., Epple, U., Hadlich, T.: Industrie 4.0 plug-and-produce for adaptable factories: example use case definition, models, and implementation. Tech. rep., Federal Ministry for Economic Affairs and Energy (BMWi) (2017)
6. Blanchet, M., Rinn, T., Von tharden, G., De Thieulloy, G.: Industry 4.0 the new industrial revolution how Europe will succeed. Think Act (2014)

7. Bohuslava, J., Martin, J., Igor, H.: TCP/IP protocol utilisation in process of dynamic control of robotic cell according Industry 4.0 concept. In: 2017 IEEE 15th International Symposium on Applied Machine Intelligence and Informatics (SAMI), pp. 000217–000222 (2017). https://doi.org/10.1109/SAMI.2017.7880306

8. Bragança, S., Costa, E., Castellucci, I., Arezes, P.M.: A brief overview of the use of collaborative robots in industry 4.0: human role and safety. In: Arezes, P., et al. (eds.) Occupational and Environmental Safety and Health. Studies in Systems, Decision and Control, vol. 202. Springer, Cham (2019). https://doi.org/10.1007/978-3-030-14730-3_68

9. Pérez, J.D.C., Buitrón, R.E.C., Melo, J.I.G.: Methodology for the retrofitting of manufacturing resources for migration of SME towards Industry 4.0. In: Florez, H., Diaz, C., Chavarriaga, J. (eds.) Applied Informatics, pp. 337–351. Springer, Cham (2018)

10. Gonzalez, A.G.C., Alves, M.V.S., Viana, G.S., Carvalho, L.K., Basilio, J.C.: Supervisory control-based navigation architecture: a new framework for autonomous robots in Industry 4.0 environments. IEEE Trans. Ind. Inform. **14**(4), 1732–1743 (2018). https://doi.org/10.1109/TII.2017.2788079

11. Hermann, M., Pentek, T., Otto, B.: Design principles for Industrie 4.0 scenarios: a literature review (2015)

12. MacDougall, W.: Industrie 4.0 smart manufacturing for the future. Tech. rep., Germany Trade & Invest (2014)

13. Mendes, M.J.G.C., Neto, M.M.S., Calado, J.M.F.: Fault diagnosis system via internet applied to a gantry robot—a proposal for Industry 4.0. In: 2018 IEEE International Conference on Autonomous Robot Systems and Competitions (ICARSC), pp. 160–166 (2018). https://doi.org/10.1109/ICARSC.2018.8374177

14. VDMA: Industrie 4.0 communication guideline - based on OPC UA. Tech. rep., VDMA Industrie 4.0 Forum (2017). https://industrie40.vdma.org/documents/4214230/20743172/Leitfaden_OPC_UA_Englisch_1506415735965.pdf/a2181ec7-a325-44c0-99d2-7332480de281

15. Weyer, S., Schmitt, M., Ohmer, M., Gorecky, D.: Towards Industry 4.0 - standardization as the crucial challenge for highly modular, multi-vendor production systems. Int. Fed. Autom. Control **48**(3), 579–584 (2015)

16. Wruetz, T., Golz, J., Biesenbach, R.: A wireless multi-robot network approach for Industry 4.0 using RoBO2L. In: 2017 14th International Multi-Conference on Systems, Signals Devices (SSD), pp. 661–664 (2017). https://doi.org/10.1109/SSD.2017.8166965

Automation of a Test Bench for Aluminum Anodizing

Hernando González$^{(\boxtimes)}$, Luis Reatiga, Carlos Arizmendi,
and Pablo Muñoz

Universidad Autónoma de Bucaramanga, Bucaramanga, Colombia
{hgonzalez7, ljaimes9, carizmendi,
jmunoz435}@unab.edu.co

Abstract. The paper presents the design of a test bench to perform electro-chemical process of anodizing on an aluminum piece. The objective is to ana-lyze the influence of the electrolytic solution, temperature and current in the anodic layer formation. The temperature and current controllers are imple-mented in an open hardware Arduino technology, which guaranteeing the sta-bility of these variables in the process. A mechanical agitator was also added to maintain the homogeneous solution in the tests. An user interface is designed in LabView, where the variables are monitored in real time and allow the storage of the historical data for later analysis.

Keywords: Anodizing process · Temperature controller · Current controller · PID control · Robust control

1 Introduction

The electrochemical cells are a device used for the decomposition of ionized substances by means of the electric current. These devices have an anode and cathode which are submerged in an electrolytic solution, basically acids. Anode is the electrode where oxidation occurs and cathode is the electrode where reduction occurs. These cells have several applications in industry, one of them is aluminum anodizing process in which, the metal surface is covered with a decorative, durable and corrosion resistant layer. Aluminum is ideal for anodizing, although other non-ferrous metals, such as magne-sium and titanium, can also be anodized [1]. In the elaboration of this process different acids can be used as electrolytic solutions such as sulphuric acid, chromic acid, oxalic acid, phosphoric acid, boric acid. In the state of the art several research's have focused on determining experimental models that relate the temperature, current and the growth of the anodic layer, working different electrolytic solution [2–5].

2 Mechanical Design

For the design of the test bench was considered an easy operation and maintenance, the following points were taken into account:

A. Martínez et al. (Eds.): LACAR 2019, LNNS 112, pp. 13–21, 2020.
https://doi.org/10.1007/978-3-030-40309-6_2

- User should adjust the distance between electrodes in order to carry out several experiments at different distances.
- Design a corrosion resistant tank in order to carry out the anodizing process.
- Pour the electrolyte solution inside the tank.

The anodizing bench has a mechanism to vary the distance between cathode and the anode. The tank was manufactured in stainless steel; dimensions are 40 cm × 30 cm × 30 cm. The didactic module has a power source of 12 v, to 5 A. Figure 1 shows up the bench.

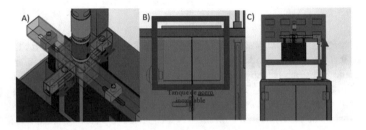

Fig. 1. Design of module for anodizing process

3   Temperature Control System

For temperature control system of the solution a 3 KW electrical resistance was implemented. This value is determined by equalizing the electric power with the caloric power delivered, Eq. (1), in which Q_R is defined as the energy consumed to make a change of 10 °C in the water temperature in three minutes, V is volume of water to be heated (12 L), ρ is the water density (998.29 Kg/m^3) and Cp is heat capacity of water (0.4270 K/Kg°C).

$$\frac{Qresistance}{t_{límit}} = \frac{M * Cp * \Delta T}{t_{límit}} = \frac{V * \rho * Cp * \Delta T}{t_{límit}} \tag{1}$$

The electrical power in the resistance is determined by a PWM signal (0% to 100%), which establishes the activation time of a relay that regulates the voltage supplied to the resistor. Identification of temperature process was performed using a first order model. The mathematical expression that identifies a first order model is presented in Eq. (2).

$$G(s) = \frac{K_T e^{-t_d s}}{\tau s + 1} \tag{2}$$

where K is the ratio between the temperature in the solution and the signal PWM in steady state, τ is time constant and T_d is delay of system. To identify the parameters of the transfer function the identification toolbox of Matlab was used. A step input is applied and the transient response of the temperature is evaluated. The parameters were: $K = 7.0504$, $T_d = 6\,s$ and $\tau = 17773\,s$.

$$G(s) = \frac{7.0504e^{-6s}}{17773\,s + 1} \tag{3}$$

Once the plant has been identified, a controller is designed to ensure the stability of the process temperature. PID control is widely used in the industry to control thermal systems with an excellent result. Equation 4 represents the mathematical expression in continuous of a PID controller.

$$u(t) = K_p\left[e(t) + \frac{1}{T_i}\int_0^t e(t)dt + T_d\frac{de(t)}{dt}\right] \tag{4}$$

where the variable $e(t)$ is the error between the set point and the output of process [6]. The PID control makes a weighting of $e(t)$, the integral of $e(t)$ and the derivative of $e(t)$: K_p makes the controller robust against possible noises, T_i reduces the error in steady state to zero and T_d makes the rise time of the controller faster; $u(t)$ is the control signal generated by the PID controller that will be applied to the process. The transfer function resulting from the controller is (5), in which $K_d = K_pT_d$ and $K_i = K_p/T_i$.

$$\frac{U(s)}{E(s)} = \frac{K_ds^2 + K_ps + K_I}{s} \tag{5}$$

Equation 5 is an improper function because the order of denominator is less than numerator order. This is mainly due to the term of the derivate of the error signal, K_ds. To solve this problem the term derivate can be approximated by a transfer function of the first order, as presented in (6). The expression approximates the derivative at low frequencies but the gain is limited by N pole to high frequencies. The transfer function of PID controller is rewritten as (7).

$$K_ds \cong \frac{K_ds}{1 + \frac{K_ds}{N}} \tag{6}$$

$$\frac{U(s)}{E(s)} = \frac{(K_dN + K_d)s^2 + (K_pN + K_IK_d)s + K_IN}{(K_ds + 1)s} \tag{7}$$

To tune the controller, the pole placement technique was used [7]. The integral component was suppressing for obtain a setting time short and a low overshoot. The PD controller has a steady state error of 0.4% and settling time of 262 s.

$$C(s) = \frac{4.5(s + 0.7)}{s + 0.1} \tag{8}$$

4 Current Control System

To regulate the current supplied between the anode and the cathode of the electro-chemical cell, a DC/DC buck converter was design (Fig. 2). Buck converter is one of the simplest but most useful power converters. It is a step-up converter that converts an unregulated DC input voltage to a regulated dc output. Figure 2 shows the basic circuit configuration used in the buck converter. As can be seen, it consists of a power MOSFET switch Q, flywheel diode D, inductor L, output capacitor C, and load resistance R. The inductor L acts as energy storage element that keeps the current flowing while the diode facilitates inductor current wheeling during the OFF time of the MOSFET. Filter made of capacitor (C) is normally added to the output of the converter to reduce output voltage ripple. Table 1 show the parameters assumed for determine the value of the inductance and the capacitor through (9) and (10).

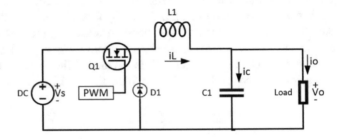

Fig. 2. Equivalent circuit of DC/DC buck converter

$$L = \frac{V_o(V_s - V_o)}{f\Delta I V_s} \tag{9}$$

$$C = \frac{V_o(V_s - V_o)}{8L\Delta V_o f^2 V_s} \tag{10}$$

Table 1. Parameters for design of DC/DC buck converter

Parameters	Value	Parameters	Value
Input voltage, V_S	12 V	ΔI	0.01 A
Commuting frequency, f	500 Hz	ΔV_o	0.01 V
Load resistance, R	1,7 Ω–7 Ω	L	0.583 H
Output current, i_O	0.01 A–10 A	C	0.00025 F
Output voltage, V_O	1 V–7 V		

The switching period is defined as the sum of the on and off intervals $T = T_{on} + T_{off}$. The ratio of the T_{on} interval to the switch period is known as the duty ratio or duty cycle $\mu = T_{on}/T$. The output voltage can be computed in terms of duty cycle during its operation in steady state. The output voltage produced by the DC-Dc buck converter is always lower as compared with the input voltage owing to its configuration. The required out voltage is controlled by varying the on time or by varying the duty cycle. The differential equation of the converter in both the modes can be obtained by using KVL and KCL [8–10]. The analysis starts from equations during the transistor's on and off states and uses an averaging method to linearize them. When the switching device is turned on, it conducts for a ratio μ of a period. When the switching device is turned off, the diode conducts for a ratio of $(1 - \mu)$ of a period. Because on and off states are presented for μ and $1 - \mu$ duration, they can be averaged by the conduction ratio. The differential equations of the converter are:

$$\frac{di_L}{dt} = -\frac{V_o}{L} + \frac{V_{in}}{L}\mu \tag{11}$$

$$\frac{dV_o}{dt} = -\frac{V_o}{RC} + \frac{i_L}{C} \tag{12}$$

The transfer function that relates the variations in the output current i_L with respect to μ is given by (13).

$$G(s) = \frac{\frac{V_{in}s}{L} + \frac{V_{in}}{LRC}}{s^2 + \frac{1}{RC}s + \frac{1}{LC}} \tag{13}$$

4.1 Quantitative Feedback Control (QFT)

Quantitative feedback control (QFT) theory is a control engineering method that explicitly proposes the use of feedback to simultaneously reduce the effects of plant uncertainty and satisfy desired performance specifications [11]. Figure 3 shows the control structure where P(s) is the transfer function of the plant, G(s) is the controller, F(s) is the prefilter; $D_1(s)$, $D_2(s)$ and $N(s)$ are disturbance signals for the system.

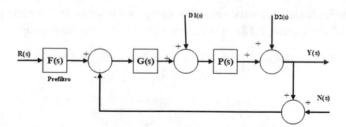

Fig. 3. QFT generic control structure

For the design of the current controller of the electrochemical cell is necessary to identify the uncertainty space of the load resistance: its nominal value is 3.98 Ω and with a 20% of variation, resulting in an interval of 3.184 Ω to 4.776 Ω. For the controller design, multiple transfer functions are generated and analyzed in a range of frequencies between 0.01 Hz to 100 Hz. Three performance specifications are defined for the design QFT controller, $G(s)$, which are defined in (14), (15) and (16). The first function characterizes the rejection of noise picked up by the sensor, the second and third functions correspond to the effect of disturbance signals on the variable to be controlled. Equation (17) is for the design of the prefilter, these transfer functions correspond to the form of the transient response of the output variable before a single step input. Normally, the lower limit is characterized by an underdamped response and the upper limit by an overdamped response. In order to resolve these inequalities, the following parameters must be quantified $\delta_u(\omega)$, $\delta_s(\omega)$, $\delta_e(\omega)$, $\alpha_L(\omega)$ y $\alpha_u(\omega)$, either as constants or from transfer functions representing the desired plant dynamics. The robust stability criterion is defined of (18) and (19), with $\delta_u(\omega) = 1.3$ resulting in a 45° phase margin and 5 dB gain margin [12, 13]

$$\left| \frac{Y(j\omega)}{R(j\omega)F(j\omega)} \right| = \left| \frac{P(j\omega)G(j\omega)}{1 + P(j\omega)G(j\omega)} \right| \leq \delta_T(\omega) \tag{14}$$

$$\left| \frac{Y(j\omega)}{D_2(j\omega)} \right| = \left| \frac{1}{1 + P(j\omega)G(j\omega)} \right| \leq \delta_s(\omega) \tag{15}$$

$$\left| \frac{U(j\omega)}{R(j\omega)F(j\omega)} \right| = \left| \frac{G(j\omega)}{1 + P(j\omega)G(j\omega)} \right| \leq \delta_e(\omega) \tag{16}$$

$$\alpha_L(\omega) \leq \left| F(j\omega) * \frac{P(j\omega)G(j\omega)}{1 + P(j\omega)G(j\omega)} \right| \leq \alpha_U(\omega) \tag{17}$$

$$MF \geq 180° - \arccos\left(\frac{0.5}{\delta_r^2} - 1 \right) \tag{18}$$

$$MG \geq 1 + \frac{1}{\delta_r} \tag{19}$$

The transfer functions for the controller design were defined taking into account the bandwidth of the system, 0.18 Hz, obtaining the following relationships:

$$\delta_T(\omega) = \frac{s^2}{s^2 + 1.333s + 0.8883} \tag{20}$$

$$\delta_s(\omega) = \frac{0.4777}{s^2 + 0.9774s + 0.4777} \tag{21}$$

$$\delta_u(\omega) = \frac{s^2}{s^2 + 1.066s + 0.5685} \tag{22}$$

$$\alpha_U(\omega) = \frac{0.9s + 0.09}{s^2 + 0.78s + 0.09} \tag{23}$$

$$\alpha_L(\omega) = \frac{0.1299}{s + 0.1299} \tag{24}$$

Using the loop-shaping technique, a G(jω) controller is introduced that manipulates the closed-loop function until it meets the restrictions imposed by the performance specifications. The closed-loop response at the frequency of interest must be above the intercept of the contours at each frequency of interest. This is achieved by adding poles and zeros to the closed-loop function until the desired response is achieved. The transfer function of the designed controller is shown in (25). In the design of the prefilter, the solution of the quadratic inequality was taken as a reference, thus, the response in the frequency of the plant should be found between the upper limit and lower limit of the functions $\alpha_U(\omega)$ y $\alpha_L(\omega)$. The pre-filter allows the behavior of the closed loop to be modeled at each frequency according to the robust specification required for signal tracking. The pre-filter has been designed for the system $F(j\omega)$ presented in (26).

$$G(j\omega) = \frac{0.4469s + 1}{0.02778s^2 + s} \tag{25}$$

$$F(j\omega) = \frac{0.05869s + 1}{0.0004936s + 1} \tag{26}$$

5   Experimental Validation

An interface was designed in Labview which communicates with an open hardware Arduino technology. The Arduino measures the temperature and current variables, in addition to generating the control signals. The controllers were discretized with a sampling period of 0.025 s. In the LabVIEW interface, the user can define the setpoint values and store the historical values of each variable (Fig. 4).

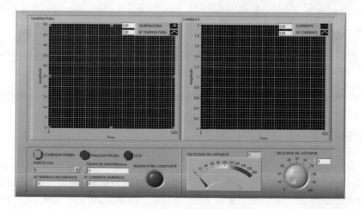

Fig. 4. Interface in Labview

Figure 5(a) shows the transient response of the temperature controller for a variable setpoint, in a range, from room temperature to 35 °C. This controller maintaining the temperature in steady state with an error of 1%. Figure 5(b) shows the transient response of the current controller for the setpoint between 0 A to 1.5 A; it is over-damped and the settling time is less than 5 s. In order to validate the test bench, an aluminum piece is anodized with sulfuric acid, the concentration is 15%, the current is fixed to 0.6 A, ambient temperature and a stirring speed of 30 RPM. The piece has an area of 33 cm^2 and the process has a duration of 6000 s. Once the process was finished, the piece was subjected to a microscopy test in order to make a measurement of the anodic layer created. Figure 6 shows up the photo taken to measure the anodic layer, reaching a thickness between 20.18 μm to 21.16 μm. Once the anodic layer was measured, a chemical analysis was carried out, confirming the existence of oxygen and aluminum in the analyzed layer (Al_2O_3 – alumina layer).

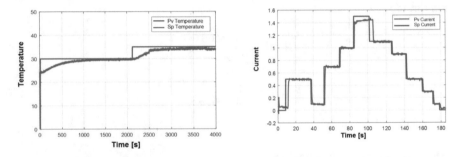

Fig. 5. Transient response. (a) Temperature controller (b) Current controller

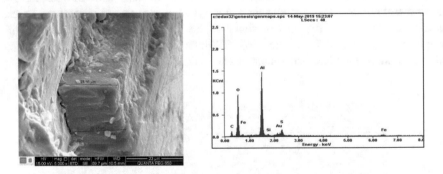

Fig. 6. Aluminum anodizing process (a) Anodic layer (b) Chemical analysis

6   Conclusion

An automatic control system for the electrochemical cell on the test bench was developed and experimentally validated. The developed system has the capacity to store automatically the information corresponding to the tests, to analyze the obtained data and to facilitate the development of new investigations. The PD controller implemented for the temperature of the solution presents a good transient response. Similarly, the current controller presents an excellent response and stability during the process, in which the load resistance changes as a function of time. Guaranteed the stability of the temperature and the current, a process of uniform anodizing is achieved in the aluminum pieces, with a growth of the layer upper 20 µm.

References

1. Paredes, J.L.: El uso de la anodizacion en materiales. Casa abierta al tiempo **IV**(28), 59–65 (2010)
2. Aviles, A., Vinueza, K., Galo, D.: Diseño e implementacion de un sistema de control para anodizado tipo 2. Sangolqui, Universidad de las Fuerzas Armadas, Departamento de Electrica y Electronica (2017)
3. Blanco, C.: Evaluacion del efecto de la temperatura en el anodizado del aluminio de alta pureza en acido sulfurico para su potencial aplicacion biomedica. Universidad Industrial de Santander Facultad de Ingenierias Fisica-Quimicas, Bucaramanga (2011)
4. Vergara, A.F.: Influencia de las variables de operacion del proceso de anodizado del aluminio sobre el comportamiento anticorrosivo de la pelicula de oxido de aluminio. Universidad Nacional de Ingenieria Facultad de Ingenieria Quimica y Textil, Lima (2010)
5. Aperador, W., Vera, E., Ipaz, L.M.: Efecto de la densidad de corriente sobre la morfologia y las propiedades electroquimicas en peliculas anodicas porosas, crecidas sobre AA 2024-T3. Rev. EIA **15**, 9–19 (2011)
6. Ogata, K.: Modern Control Engineering, 5th edn. Prentice Hall, New Jersey (2010)
7. Marin, L., Alfaro, V.: Sintonización de controladores por ubicación de polos y ceros, San Jose. IEEE CONESCAPAN XXVI, Costa Rica (2007)
8. Rashid, M.H.: Electronica de Potencia. Prentice Hall hispanoamericana, S.A, Naucalpan de Juarez. Edo.de Mexico (1995)
9. Sira Ramirez, H., Silva Ortigoza, R.: Control Design Techniques in Power Electronics Devices. Springer, Cham (2006). ISBN 978-1-84628-459-5
10. Suntio, T., Messo, T., Puukko, J.: Power Electronic Converters: Dynamics and Control in Conventional and Renewable Energy Applications. Wiley-VCH, Weinheim (2017). ISBN 978-3-527-34022-4
11. Elso, J., Monserrat, G., Garcia, M.: Quantitative feedback control for multivariable model matching and disturbance rejection. Int. J. Robust Nonlinear Control **27**(1), 121–134 (2017)
12. Chait, Y.: Optimal automatic loop-shaping of QFT controllers via convex optimization. In: Proceedings Symposium on Quantitative Feedback Theory and other Frequency Domain Methods and Applications, Glasgow, Scotland, pp. 13–28 (September 1997)
13. Chait, Y., Borghesani, C., Zheng, Y.: Single-loop QFT design for robust performance in the presence of non-parametric uncertainties. ASME J. Dyn. Syst. Meas. Control **117**, 420–425 (1995)

Towards Automatic UAV Path Planning in Agriculture Oversight Activities

Daniel Palomino-Suarez and Alexander Pérez-Ruiz[✉]

Escuela Colombiana de Ingeniería Julio Garavito, Bogotá D.C., Colombia
daniel.palomino@mail.escuelaing.edu.co,
alexander.perez@escuelaing.edu.co

Abstract. By 2050, growth population worldwide will demand a great amount of resources, specially in agricultural sector. In order to supply such resources, farmers will need advance tools and techniques to improve the efficiency along all farming processes e.g. precision farming, Unmanned Vehicles Systems (*UVS*). This paper shows the simulation results of an algorithm developed to perform the path planning process for Unmanned Aerial Vehicles (*UAVs*) autonomously, in agricultural environments. The aim of this project is to automate such process and provide the appropriate conditions to execute further supervision activities. The algorithm takes into account photogrammetric parameters such as the ground sample distance (*GSD*) and the overlap between photos. Image processing techniques are implemented to identify the crop boundary and rows' orientation. The project was developed using open-source tools such as OpenCV, ROS and Gazebo. The paths generated by this algorithm allow the UAV not only to go across the identified crop following the row orientation, but also to guarantee the desired resolution, specified by the *GSD*.

Keywords: Path planning · Unmanned Aerial Vehicles · Photogrammetry · Image processing · Precision agriculture · ROS

1 Introduction

According to projections presented by [20], worldwide population could increase until 9.7 billion people by 2050. Such scenario will demand lots of resources from agricultural sector in order to fulfil people's food needs. Therefore, new techniques and tools need to be developed to improve the efficiency along each process executed by farmers to take advantage of the available resources as much as possible. This has drawn attention of researchers and significant developments have been made in fields like Precision agriculture, remote sensing, and even Internet of things (IOT).

These disciplines use specialized tools to retrieve data from the fields, then decisions are made based on the analysis results, for example [21] develop a system to retrieve data from sensors distributed along the field with the help

A. Martínez et al. (Eds.): LACAR 2019, LNNS 112, pp. 22–30, 2020.
https://doi.org/10.1007/978-3-030-40309-6_3

of an UAV. Unmanned vehicle systems, specifically UAVs, have become key tools to perform supervision and monitoring activities, [2,15,16,18] show some remarkable contributions where UAVs are used along with RGB cameras, multi-spectral cameras and other specialized sensors to perform these activities.

However, these vehicles are dependent on whether the workspace is known or not. In the first case, efficient and customized plans can be generated using techniques described by [5,6,12]. But if the workspace is unknown, this should be identified and then the appropriate paths should be generated, [14] proposed an algorithm to perform the path planning process when the workspace dimensions are known. This points out how important the path planning process to perform further supervision activities is.

Through this paper an algorithm to automate not only the workspace identification but also the path planning process is proposed. Section 2 describes the methodology and tools used to develop the algorithm, in Sect. 3 simulation results generated by the algorithm are shown and discussed and finally in Sect. 4 conclusions and further work are discussed.

2   Methodology

In order to develop and test the algorithm a simulation environment was created using ROS (Robot Operating system) and Gazebo, which have become essential tools in the development of robotics projects. On the first hand, ROS offers multiple communication structures that allow to control robots and vehicles such as UAVs in an efficient way, on the other hand, Gazebo offers a physics engine that allow developers to simulate objects and sensors behavior. In this project the Hector Quadrotor model [11] was implemented along with simulated GPS, IMU and camera as a main sensors. Crops were created using a 3D modeller software called Blender, another OpenSource tool, Fig. 1 shows an example of the simulation environment.

With this environment created, the algorithm was developed in two main stages. First, in the Crop features identification stage, the UAV takes off until the maximum height allowed by regulation (500 ft), where a picture is taken, and uses the image processing techniques implemented to identify the orientation of the rows as well as the crop contour. Then, the path generation is executed. In this stage, transformations between different frames are executed and strong geodesic concepts are used to generate the way-points across the identified crop that will guarantee the overlap and *GSD* specified as input parameters.

2.1   Crop Features Identification

When the UAV has reached the maximum height a picture is taken and then processed. This algorithm will be used on images taken with RGB sensor or with multi-spectral camera as well. First of all, a segmentation based on color is done to identify the rows. This segmentation is accomplished by transforming the RGB color space to the HSV color space and only those pixels whose Hue

(a)   **(b)**

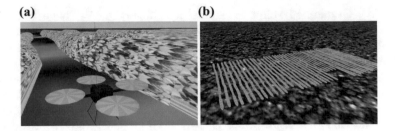

Fig. 1. Simulation environment. (a) Hector quadrotor. (b) Crop

channel value correspond to green color are accepted. This mask is used in both, row orientation and contour identification processes.

To obtain the orientation of the rows some morphological operations (erode and dilate) are performed in an iterative manner until the skeleton of the rows is obtained and then, the Hough transform for lines is used to identify the rows, as is done by [9,13,19]. The orientation of each row is found in the image frame and then, the mode is computed to be chosen as the orientation that will be followed by the UAV.

Finally, to find the crop boundary, morphological operations are executed over color-segmented mask in such a way that each row is expanded until it is overlapped with the surrounding rows. Then the contour can be easily identified and defined as a sequence of points in the image frame. It is worthed to mention that the previous processes were implemented as a node in ROS using the OpenCV library, which has implemented multiple image processing techniques. Figure 2 shows part of the image processing and final results of the crop features identification.

(a)   **(b)**   **(c)**

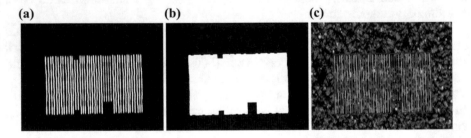

Fig. 2. Features Crop identification. (a) Color-based mask. (b) Boundary. (c) Image processing results.

2.2   Way-Points Generation and Georeferencing Process

The results from previous stage are given in the image frame and since there are multiple frames involved as shown in Fig. 3, before the generation of the way-points, it is necessary to apply some transformations between them. These transformations will use the information provided by the GPS, IMU sensors, and the camera parameters when picture is taken, as described by [1,7,8,22]. The purpose of using the transformations is to manage all the information in the navigation or mapping frame, because it is easier to plan the trajectories over a Cartesian coordinate system.

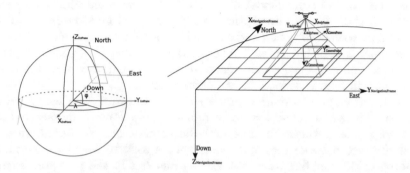

Fig. 3. Involved frames: camera frame, image frame, body frame, ECEF frame, navigation/mapping frame and geodetic frame

When a photogrammetric process needs to be done, some of the most relevant parameters are the *GSD*, sidelap, and endlap, which in this algorithm are provided as inputs. The *GSD* (1) defines the spatial resolution per pixel, while endlap and sidelap define the frontal and lateral overlap between adjacent photos [4], respectively. These parameters are used to compute the UAV flight height, the distance between each photography, commonly known as base (2), and the distance between each flight line or line distance (3).

$$GSD = \frac{FH \times SW}{FL \times IW} \tag{1}$$

$$Base = GSD \times IH \times \left[\frac{100 - EL}{100}\right] \tag{2}$$

$$LineDistance = GSD \times IW \times \left[\frac{100 - SL}{100}\right] \tag{3}$$

Where FH is FlightHeight, SW is SensorWidth, FL is FocalLength, SL is Sidelap, EL is Endlap IH is ImageHeight and IW is ImageWidth. Both Endlap and Sidelap are represented in percentage value.

With the crop features information in the navigation frame, the way-points can be generated. First the height that will guarantee the provided *GSD* needs to be computed. This can be done by using the *GSD* definition in Eq. (1), which depends on the image resolution, sensor size, focal length and flight height.

Then, the initial point needs to be identified. This is done by creating a line that contains the crop's centroid and whose director vector is parallel to the row orientation vector, then the distance from the line to each point that defines the contour is computed. The point whose distance to the line is the largest will be chosen as the initial point.

To identify the number of lines that need to be generated across the field, as well as the UAV's lateral movement direction an iterative process is executed. A line that contains the initial point and whose director vector is parallel to the row orientation vector is created, then the line is moved an amount equal the line distance (3) in perpendicular direction to the row orientation vector and it is verified if there is any interception. If true, this will give the information of the lateral movement direction and also will point that the number of lines to be generated needs to be increased. This process is executed until none interception is found and an additional line is added in order to guarantee that the borders of the crop will appear at least in two photographies.

Finally, through an iterative process the navigation frame way-points are generated. The same line described in previous step is used. In each iteration, the interceptions between the line and the contour are computed, then, the largest distance between the found intercepts is used to calculate the number of photographies per line using the Eq. (4) and finally the way-points can be generated. Once the way-points in navigation frame have been generated, some transformations are performed to obtain the GPS coordinates that UAV will use to go across the crop. Algorithm 1 summarizes the processes that are done in the path planing generation.

$$NumberOfPhotographies = \frac{LineLength}{Base} \tag{4}$$

All this development has been made inside the ROS concepts, with an appropriate net of nodes using publisher–subscriber and actions as well.

3   Results

The algorithm was tested using multiple crop configurations. In each of them changes in the input parameters, the crop shape, the row orientation with respect to the crop geometry and the crop orientation with respect to the camera frame were made. Figure 4 shows generated paths using some of the tested crop configurations. In each image the red lines represent the crop boundary, while blue and green connected points represent the generated paths.

In Fig. 4a, where crop has a rectangular shape and is horizontally aligned with respect to the camera frame, can be noticed that the generated paths cover only the region of interest. This is done not only when row orientation is aligned with the crop shape (green path), but also when it is not (blue path). Next, in Fig. 4b the crop has a rectangular shape, but is not horizontally aligned in relation to the camera frame. However, just like in Fig. 4a, most of the paths' points cover only the region of interest and rows' orientation does not represent any problem. In this case, the green path is generated for a crop whose row orientation is aligned in relation to its geometry, while the blue one does not follow this condition.

Algorithm 1: Automatic Path-planning algorithm

Input: GSD, endlap,sidelap
Result: Path Plan(Waypoints_GPS)
executeTakeoff;
$image = takePicture$;
$mask = applyColorSegmetation(image)$;
$skeleton = erode(mask)$;
while $!skeletonFound$ **do**
   | $skeleton = applyMorpholgycalMperations(skeleton)$;
   | $skeletonFound = validateSkelton$;
end
$lines = findHoughLines(skeleton)$;
$rowOrientationVector_{IF} = mode(orientation(lines))$;
$countour_{IF} = applyMorpholgycalOperations(mask)$;
while $!countourFound$ **do**
   | $countour_{IF} = applyMorpholgycalOperations(countour)$;
   | $countourFound = validateCountour$
end
$countour_{NAVF} = applyIF2NF\_Transform(countour_{IF}, IMU, GPS)$;
$rowOrientationVector_{NAVF} =$
$applyIF2NAVF\_Transform(rowOrientationVector_{IF}, IMU, GPS)$;
$FlightHeight = compute\_flight\_height(GSD)$;
$Base = compute\_base(endlap)$;
$LineDistance = compute\_LineDistance(SideLap)$;
$InitialPoint =$
$findInitialPoint(countour_{NAVF}, rowOrientationVector_{NAVF})$;
$[NumberOfLines, LateralMovementVector] =$
$findNOL\_LMV(countour_{NAVF}, rowOrientationVector_{NAVF})$;
for $i \leftarrow 0$ **to** $NumberOfLines$ **do**
   | $Intercepts =$
   | $findIntercepts(rowOrientationVector_{NAVF}, countour_{NAVF})$;
   | $LineLength = compute\_LineLength(Intercepts)$;
   | $NumberOfPhotos = compute\_NumberOfPhotos(Base, LineLength)$;
   | $Waypoints_{navf}[i] =$
   | $generate\_wayopitns_{NAVF}(NumberOfPhotos, InitialPoint)$;
   | $InitialPoint = updateInitialPoint(Waypoints_{NAVF}, countour_{NAVF})$;
end
$Waypoints_{GPS} = applyNF2GeodF\_Transform(Waypoints_{NAVF}, IMU, GPS)$

Finally, in Fig. 4c, even though the *GSD* was reduced in relation to Figs. 4a and 4b, and a none rectangular crop shape was used, appropriate paths are generated by the algorithm. Both, blue and green paths cover the area of interest in an efficient way regardless the crop shape.

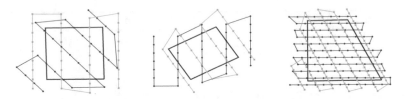

Fig. 4. Simulation results varying the alignment between camera, field and crop.

The generated paths were successfully executed by the UAV in the simulation environment, but it was noticed that is necessary to enhance the control strategies that manage the UAV's movements. However, the main purpose at this work, which is the path planning, was accomplished.

4   Conclusions and Future Work

An algorithm that automates the work-flow related to the path planning process was generated. This algorithm has the capability to identify crops based on color features, it is not dependent on the crop geometry and allows the UAV to execute the plan following the direction of the rows. The paths generated by this algorithm fulfill photogrammetric requirements, specifically those related to the overlap between photos and *GSD*. These features will allow farmers or researchers to perform further and more advance tasks, like those mentioned by [14, 16, 17].

Since the algorithm was developed in multiple ROS nodes, embedded systems, compatible with ROS, can be used to execute the algorithm and work along with the autopilot system of real UAVs. However, slight changes can be done to the algorithm in order to discard some points that are outside crop's boundaries and do not provide relevant information.

The crop identification process can be improved by using multi-spectral data as is done by [3, 10, 17]. Multi-spectral data can be used to estimate more relevant information related to crops or plants. For example, vegetation indexes like Normalized difference vegetation index (NDVI) could be used to determine whether vegetation is present in a zone or not. Unfortunately, this cannot be done in simulation environment, because neither ROS nor Gazebo have implemented a multi-spectral camera sensor yet.

With the results provided by the algorithm, high-quality image-data can be acquired to perform tasks such as image-mosaicking [14]. This will help researchers and farmers to manage crops more efficiently when performing supervision or monitoring activities.

This algorithm will be tested in a second stage, using an NVidia Jetson TX2 as a companion computer along with the pixhawk 2 autopilot to perform the monitoring tasks autonomously without the human intervention.

References

1. Cai, G., Chen, B.M., Lee, T.H.: Coordinate Systems and Transformations, pp. 23–34. Springer, London (2011)
2. Colomina, I., Molina, P.: Unmanned aerial systems for photogrammetry and remote sensing: a review. ISPRS J. Photogramm. Remote Sens. **92**, 79–97 (2014)
3. Comba, L., Gay, P., Primicerio, J., Aimonino, D.R.: Vineyard detection from unmanned aerial systems images. Comput. Electron. Agric. **114**, 78–87 (2015)
4. Förstner, W., Wrobel, B.P.: Photogrammetric Computer Vision: Statistics, Geometry, Orientation and Reconstruction, 1st edn. Springer, Cham (2016). Incorporated
5. Galceran, E., Carreras, M.: A survey on coverage path planning for robotics. Robot. Auton. Syst. **61**(12), 1258–1276 (2013)
6. Goerzen, C., Kong, Z., Mettler, B.: A survey of motion planning algorithms from the perspective of autonomous UAV guidance. J. Intell. Robot. Syst. **57**(1), 65 (2009)
7. Grewal, M.S., Weill, L.R., Andrews, A.P.: Global Positioning Systems, Inertial Navigation, and Integration, pp. 324–346. Wiley, Hoboken (2007)
8. Hemerly, E.M.: Automatic georeferencing of images acquired by UAV's. Int. J. Autom. Comput. **11**(4), 347–352 (2014)
9. Ji, R., Qi, L.: Crop-row detection algorithm based on random Hough transformation. Math. Comput. Model. **54**(3), 1016–1020 (2011). Mathematical and Computer Modeling in agriculture (CCTA 2010)
10. Lottes, P., Khanna, R., Pfeifer, J., Siegwart, R., Stachniss, C.: UAV-based crop and weed classification for smart farming. In: 2017 IEEE International Conference on Robotics and Automation (ICRA), pp. 3024–3031 (2017)
11. Meyer, J., Sendobry, A., Kohlbrecher, S., Klingauf, U., von Stryk, O.: Comprehensive simulation of quadrotor UAVs using ROS and Gazebo. In: 3rd International Conference on Simulation, Modeling and Programming for Autonomous Robots (SIMPAR), p. to appear (2012)
12. Sponagle, P., Salvaggio, C.: Automatic mission planning algorithms for aerial collection of imaging-specific tasks, vol. 10218 (2017)
13. Ramesh, K., Chandrika, N., Omkar, S., Meenavathi, M., Rekha, V.: Detection of rows in agricultural crop images acquired by remote sensing from a UAV. Int. J. Image Graph. Signal Process. **8**(11), 25 (2016)
14. Rojas, J., Martinez, C., Mondragon, I., Colorado, J.: Towards image mosaicking with aerial images for monitoring rice crops. In: Chang, I., Baca, J., Moreno, H.A., Carrera, I.G., Cardona, M.N. (eds.) Advances in Automation and Robotics Research in Latin America, pp. 279–296. Springer, Cham (2017)
15. Rokhmana, C.A.: The potential of UAV-based remote sensing for supporting precision agriculture in Indonesia. Procedia Environ. Sci. **24**, 245–253 (2015). The 1st International Symposium on LAPAN-IPB Satellite (LISAT) for Food Security and Environmental Monitoring
16. Roldán, J.J., del Cerro, J., Garzón-Ramos, D., Garcia-Aunon, P., Garzón, M., de León, J., Barrientos, A.: Robots in agriculture: state of art and practical experiences. In: Service Robots. IntechOpen (2017)
17. Romero, M., Luo, Y., Su, B., Fuentes, S.: Vineyard water status estimation using multispectral imagery from an UAV platform and machine learning algorithms for irrigation scheduling management. Comput. Electron. Agric. **147**, 109–117 (2018). http://www.sciencedirect.com/science/article/pii/S0168169917315533

18. Shakhatreh, H., Sawalmeh, A.H., Al-Fuqaha, A., Dou, Z., Almaita, E., Khalil, I., Othman, N.S., Khreishah, A., Guizani, M.: Unmanned aerial vehicles (UAVs): a survey on civil applications and key research challenges. IEEE Access **7**, 48572–48634 (2019)
19. Soares, G.A., Abdala, D.D., Escarpinati, M.: Plantation rows identification by means of image tiling and Hough transform. In: VISIGRAPP (4: VISAPP), pp. 453–459 (2018)
20. UN: Population   (2017).   https://www.un.org/en/sections/issues-depth/population/index.html
21. Vasisht, D., Kapetanovic, Z., Won, J., Jin, X., Chandra, R., Sinha, S., Kapoor, A., Sudarshan, M., Stratman, S.: Farmbeats: an IoT platform for data-driven agriculture. In: 14th USENIX Symposium on Networked Systems Design and Implementation (NSDI 17), pp. 515–529. USENIX Association, Boston (2017)
22. Xiang, H., Tian, L.: Method for automatic georeferencing aerial remote sensing (RS) images from an unmanned aerial vehicle (UAV) platform. Biosyst. Eng. **108**(2), 104–113 (2011)

Computer Vision for Recognition of Fruit Maturity in Amazonian Palms Using an UAV

Willintong Marín[✉], J. Colorado, and Iván Mondragón Bernal

School of Engineering, Pontificia Universidad Javeriana,
Cra. 7 #40-62, Bogota, Colombia
{willintong.marinr, coloradoj,
imondragon}@javeriana.edu.co

Abstract. This paper presents the integration of well-known computer vision methods to identify the stage of maturity of the Asai, Seje and Moriche fruits in Amazonian palm based on aerial images acquired with an Unmanned Aerial Vehicle (UAV). Despite the aim of this research is to use both multispectral (NIR) and thermal cameras to acquire imagery at different wavelengths, this paper is limited to a maturity classification by extracting features from visible spectrum imagery (VIS). We have implemented an algorithm that combines the VIS image histogram with a mask filter and the corresponding thresholding to process the acquired images. A classifier is being used to recognize the maturity stage of Moriche palm fruits. Classification results have shown an overall accuracy of 66.5% and a performance of 58.33%. These preliminary results confirm that we need to include NIR information for enabling the extraction of more relevant features related to the fruit maturity stage. An approach based on NIR vegetative indices will be implemented in upcoming work.

1 Introduction

The Amazon Scientific Research Institute SINCHI has formulated a management plan for the sustainable use of native species such as the Asai, Seje and Moriche fruit species of the Amazon region with the aim of increasing exportations of an oil-extracted product with high nutritional value. The oil extracted from these fruits has several benefits in the treatment of tuberculosis, lung diseases or respiratory problems. Due to its high copper content, it also has benefits in the formation of hemoglobin and in the development and maintenance of bones and tendons [1].

These plants are usually located in non-cultivated dense forests. An adequate management plan allows for a sustainable use of these products, maintaining the preservation conditions and income with minimal impact. These palms are in dispersed and flood zones with wetlands. In addition, the fruit can only be gathered in a certain stage of maturity, situation that makes the harvest expensive.

This work seeks to facilitate the collecting fruit process by integrating an autonomous sorting framework using Unmanned Aerial Vehicles (UAVs) and computer vision algorithms for identifying the maturity fruit stage.

This study has been conducted in Colombia (Guaviare Department in the Amazonian region) in which the Moriche palm Aguaje (*Mauritia Flexuosa L. f.*) typically

A. Martínez et al. (Eds.): LACAR 2019, LNNS 112, pp. 31–39, 2020.
https://doi.org/10.1007/978-3-030-40309-6_4

Fruit	Color Scale	Maturity status
	1	Green
	2	Pinton 1
	3	Pinton 2
	4	Pinton 3
	5	Mature

Fig. 1. Moriche maturity classification according [4]

grows 40 m in height with a ripening cycle of one year according to [2]: its fruiting occurs between March and May and each fruit takes about four months to be formed and other four months to mature [3].

Figure 1 presents a Moriche maturity classification using a stage scale presented in [4]. The first stage is the green, followed by pinton 1, pinton 2, pinton 3 and ripe.

Computer vision is a tool that has been widely employed in industry processes for fruit selection, identification and sorting. The computer vision is also useful in aerial robotics focused on precision agriculture.

Image processing on visible images combined with machine learning for maturity classification have been used for different kind of fruits classification. In the work presented in [5], authors developed a computational tool for the identification of the maturation stage of granadillas. The region of interest (ROI) of the images belonging to the fruit was extracted by the Otsu technique using OpenCV libraries in Python. Images were captured under controlled conditions and the classification was carried out through the grouping analysis, in which 110 RGB points belonging to each ripening stage of the granadilla were assigned. The results obtained showed 92.6% accuracy in the identification of the maturation stage, from a set of 90 images obtained from 90 fruits in different stages of maturation, compared to the manual analysis according to the provisions of the Colombian Technical Standard (NTC 4101).

A classification based on red histogram form visual images of peach fruits is presented in [6]. The author explains the change in the absorption of chlorophyll based on the ripeness of the fruit, demonstrating that the reflection increases given the degradation of chlorophyll. It concluded that the classification method improves when using operations between R/IR layers.

In the work presented in [7], a technique to extract characteristics of the region of interest in blueberry fruits was applied based on the theory of information using the

Kullback-Leibler divergence. They use the closest neighbor K, the support vector machine and AdaBoost as algorithms for classification. The highest yield they reached was 88%. The images used were hyperspectral and taken without lighting control.

In [8], both NIR and visible images of nectarine fruits were used to monitor the fruit's maturity by using techniques of minimal partial frames for the selection of variables. Similarly, in [5] a computational tool for the identification of the ripening state of granadillas by analyzing RGB imagery was also implemented. An image segmentation process using the Otsu technique was used. Experimental results reported an accuracy of 92.6%. Finally, in [9], an algorithm for the image processing and classification of the ripening stage of the pineapple perolera fruit based on Otsu and HSV segmentation methods achieved an accuracy of 96.36%.

According to the aforementioned literature review, we have determined that applying classical threshold segmentation and extracting features by using the histogram in the color space and the Naives Bayes algorithm could be useful for the maturity classification of Moriche.

2 Materials and Methods

The proposed methodology for palms fruit maturity recognition comprises: the capture of the images, passing through features extraction and finally, fruit classification. Images were captured by flying with an UAV near to palms without illuminance control. The captured images include both the fruits as well as leaves. Figure 2 shows the general structure of the proposed system.

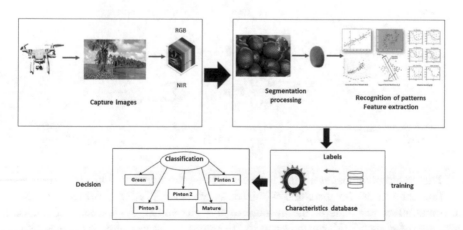

Fig. 2. General structure of project. The proposed system includes image capturing from a UAV, images processing and features extraction, ripeness classification, palms geolocation and the fruit ripeness state identification.

The SINCHI Institute has generated a Moriche fruits visible image dataset [4] that have been used for the classification process on the initial stage of this project. The dataset contains 30 images for each class. In addition, we have captured images with the proper resolution within a controlled directional light testbed.

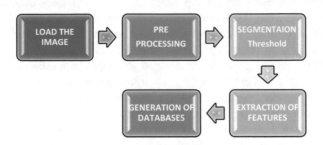

Fig. 3. Block diagram, image processing

Figure 3 presents the image processing method for maturity identifycation. Each image is pre-processed in gray scale; a Gaussian filter is applied and then segmented applying a thresholding method with the purpose of obtaining a mask filter. Equation 1 shows the match-function to obtain the thresholding, which is applied to the histogram function, passing it as a parameter. The histogram is applied exclusively in the region of interest of the original image. Then, statistical information is extracted from this region.

$$\sigma_w^2 = q_1(t)\,\sigma_1^2(t) + q_2(t)\,\sigma_2^2(t) \tag{1}$$

Where

$$q_1 = \sum_{i=1}^{t} P(i) \ \& \ q_1 = \sum_{i=t+1}^{I} P(i)$$

$$\mu_1(t) = \sum_{i=1}^{t} \frac{iP(i)}{q_1(t)} \ \& \ \mu_2(t) = \sum_{i=t+1}^{t} \frac{iP(i)}{q_2(t)}$$

$$\sigma_1^2(t) = \sum_{i=1}^{t} [i - \mu_1(t)]^2 \frac{P(i)}{q_1(t)} \ \& \ \sigma_2^2(t) = \sum_{i=t+1}^{I} [i - \mu_1(t)]^2 \frac{P(i)}{q_2(t)}$$

Due to bimodal images are used, the Otsu's algorithm tries to find a threshold value (t) that minimizes the weighted within-class variance [5]. The extracted statistical information corresponds to the average of the probability distributions of the color histogram for each RGB channel and to the grayscale image, thus, it is constructed a database with four characteristics and the corresponding label. In order to reduce complexity, only two classes were defined (green and ripe).

The classification is performed using the *Rapidminer software* [1]. Figure 4 shows the block diagram of the developed model.

---

[1] https://rapidminer.com.

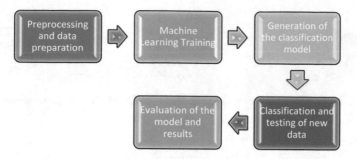

Fig. 4. Block diagram, fruit classification process of the Moriche palm

The pre-processing step allows for defining the features and the number of classes to be implemented. Identification was divided in 2 classes. The second block contains the training and cross-validation process, in which the *Rapidminer* function automatically divides the database in a proportion 70/30 for training and testing sets. The Naives Bayes algorithm is trained with 70% of the data considering the label vector of the previously defines classes, then automatically applies this model on 30% of the data as a test.

The final step is the evaluation of the generated model with the new dataset, which is supplied only with the attributes (data are not labeled). This new test dataset will evaluate algorithm performance by applying a metric based on the confusion matrix applied over applied model results.

3   Results and Discussion

Table 1 shows this characteristics table including the label of each item. This table is composed by the statistical information of each channel and the grayscale image. Four attributes were obtained for each image.

Table 1. Characteristics table for Moriche dataset.

Row No.	Media CB0	Media CB1	Media CB2	Media EG	Class
1	38.164	38.652	38.660	39.386	C1
2	46.168	45.602	46.1	46.589	C1
3	101.684	97.887	94.980	101.789	C1
4	93.781	91.648	89.273	93.804	C1
5	92.848	92.848	92.847	92.847	C1

3.1   Preprocessing Stage

The image pre-processing includes a color to gray scale transformation, allowing unifying channels and manage a single intensity per pixel. A Gaussian filter is applied to smoothing the image. Figure 5 shows the results of this process.

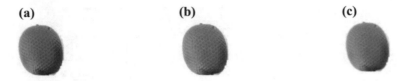

Fig. 5. Preprocessing stage. a. Original image, b. Grayscale image, c. Filtered image

3.2   Processing Stage

The image is segmented by the thresholding method. It allows obtaining a mask of the binarized image (zeros and ones). Figure 6 shows the thresholded image.

Fig. 6. Thresholdized image

3.3   Feature Extraction

The image color histograms were used to define features for the classification. Figure 7 shows a comparison between color histograms for the original image and the pre-processed image. Figure 7a corresponds to the histogram of thresholded images applying mask filter and Fig. 5b shows the histogram without this mask filter. It can be noticed a great intensity in upper limits of the histogram corresponding to white pixels of the image background.

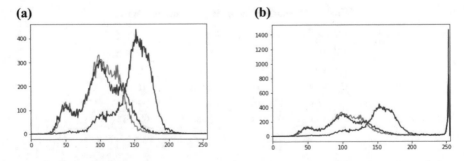

Fig. 7. Histograms of the image. a. Histogram with mask, b. Histogram without mask

3.4   Classification Process Based on Machine Learning

Prior to the classification process, the data was normalized. Table 2 shows the normalized values.

Table 2. Normalized dataset

Row No.	Class	Media CB0	Media CB1	Media CB2	Media EG
23	C1	0.658	0.657	0.657	0.567
24	C1	0.514	0.509	0.5140	0.444
25	C1	0.029	0.025	0.025	0.057
26	C2	0.078	0.074	0.074	0.098
27	C2	0.616	0.619	0.616	0.655
28	C2	0.768	0.773	0.767	0.707

Table 3 shows the statistical performance of classification process.

Table 3. Performance evaluation of Bayesian classifier for Moriche fruit.

	True C1	True C2	Class precision
Pred. C1	16.000	8.000	66.67%
Pred. C2	8.000	16.000	66.67%
Class recall	66.67%	66.67%	

As noticed in Table 3, classifier accuracy is very low with only 66.5%, for precision and 66.67% for Recall for both classes.

The next process step evaluates the generated model with the new dataset, which is supplied only with the attributes. It is expected that model classifies in green or mature each item supplied. The dataset consists of 12 fruits for the Moriche fruit. Table 4 shows the statistical information of this process.

Table 4. Statistical data of the performance evaluation of the model, for Moriche fruit.

Prediction (class)	Polynomial	0	C2 (3)	C1 (9)
Confidence (C1)	Real	0	0.105	0.948
Confidence (C2)	Real	0	0.052	0.895
Media CB0	Real	0	0.035	0.813
Media CB1	Real	0	0.036	0.812
Media CB2	Real	0	0.032	0.811
Media EG	Real	0	0.029	0.725

The statistical information obtained indicates that three fruits were classified as C2, that is, Mature and nine fruits as C1. By conducting a manual classification in the classifier, the confusion matrix presented on Table 5 is obtained. Model accuracy for this test dataset is 58.33%.

Table 5. Performance confusion matrix for test dataset.

	True C1	True C2	Precision
Predic. C1	5	1	83,33%
Predic. C2	4	2	33,33%
Recall	55,55%	66,67%	

4 Conclusions

This paper presented an integration of techniques for Moriche fruits maturity recognition based on image segmentation using thresholding and machine learning classification. Although the preliminary results are promising, it is necessary to collect additional set of images of the Moriche fruit for algorithm training and testing, than includes additional maturity stages.

The image feature extraction through histogram allowed for a basic classification of fruits maturity. Currently, improvements to this methodology are under implementation for increasing the classification accuracy by means of multispectral data. This involves features extraction through NDVI (Normalized difference Vegetation Index) over near infrared NIR and visible images (among other vegetative indices).

The low accuracy and precision readings indicated the need for a larger number of images and a greater representation of attributes, so that the algorithm training process is more robust and representative. Upcoming work is oriented towards the analysis of NIR data, improving the dataset of images, applying PCA or LDA for dimensionality reduction and using SVM and ANN for the classification of maturity based on sensitive vegetation indices.

References

1. López, R., Navarro, J.A., Montero, M.I., Vecht, K., Rodríguez, M., Polania, A.: Manual de identificación de especies no maderables del corregimiento de Tarapacá, Colombia. Bogotá D. C. (2006)
2. Hernández, J., Castro, M.S., Giraldo, S.Y., Barrera, B.: Seje, moriche, asaí: Palmas amazónicas con potencial, Primera ju. Bogotá D.C. (2018)
3. Montero, A.A., Barrera, I.M., Giraldo, J.A., Lucena, B.: Fichas Tecnicas de Especies de uso Forestal y Agroforestal de la Amazonia Colombiana. Bogotá D.C. (2016)
4. SINCHI: Fichas Palmas amazónicas con potencial Seje, Moriche y Asaí (2018)
5. Figueroa, D.E., Guerrero, E.R.: Sistema de visión artificial para la identificación del estado de madurez de frutas (granadilla). Redes Ing. **7**(1), 78 (2016)
6. Lleó, L., Barreiro, P., Ruiz-Altisent, M., Herrero, A.: Multispectral images of peach related to firmness and maturity at harvest. J. Food Eng. **93**(2), 229–235 (2009)
7. Yang, C., Lee, W.S., Gader, P.: Hyperspectral band selection for detecting different blueberry fruit maturity stages. Comput. Electron. Agric. **109**, 23–31 (2014)

8. Munera, S., Amigo, J.M., Blasco, J., Cubero, S., Talens, P., Aleixos, N.: Ripeness monitoring of two cultivars of nectarine using VIS-NIR hyperspectral reflectance imaging. J. Food Eng. **214**, 29–39 (2017)
9. Silva, L.: Su Variedad Perolera Mediante Técnicas De Visión Artificial **9**, 31–41 (2012)

Graphical Optimization for a Parallel Robot Rotation Based on Platform Initial Orientation

Alexandre Campos[1] and Yuri Daniel Moratelli[2(✉)]

[1] Universidade do Estado de Santa Catarina (UDESC), Florianópolis, SC, Brazil
alexandre.campos@udesc.br
[2] Universidade Comunitária da Região de Chapecó (UNOCHAPECÓ),
Anjo da Guarda Street, 295-D - Efapi, Chapecó, SC, Brazil
yurimoratelli@unochapeco.edu.br

Abstract. A robot platform maximum rotation is function of its initial orientation. Thus, maximizing the platform rotation depends on achieving the suitable initial orientation. This work presents a graphical optimization method that achieves this convenient orientation. The parallel robot workspace presents different rotation amplitudes around different axis. If a suitable initial orientation is chosen then the original tool orientation must be compensated, since the tool is fixed to the platform. This compensation can be performed analysing robot kinematics, singularities and collision. In singularities, robots gain or lose degrees of freedom, which makes it inoperative, or uncontrollable. The kinematic analysis is performed based on robot singularity proximity index. Therefore, this work presents a method that enables determining robot platform convenient initial orientation allowing larger rotation amplitudes around an axis.

1 Introduction

The end effector position, relative to the platform, affects the parallel robot [14]. According workspace researches, it is noticed that robot workspace is not symmetrical in all directions; some authors use circles or spheres to determine operating safe area [9,12,16]. In parallel robots, platform initial orientation influences its maximum rotation around one or more axes. This work presents a graphical optimization method to determine this convenient initial orientation, through workspace representations.

Optimization problems with up to two variables may be solved graphically [1,4]. This optimization method considers the influences of design variables in objective function value, using a visual analysis. This optimization method has the disadvantage of considering finite ranges for the design variable values. However, fortunately, in the orientation workspace representations, the variables range repeat after complete rotations.

A. Martínez et al. (Eds.): LACAR 2019, LNNS 112, pp. 40–50, 2020.
https://doi.org/10.1007/978-3-030-40309-6_5

The present work proposes a method of workspace optimization for parallel robots. This method considers the initial orientation change as a maximizing parameter, allowing robots to perform maximal rotations around certain axis without topological modifications. Additionally, it is necessary to compensate the tool orientation in relation to robot platform and platform orientation in relation to global coordinates system.

2 Concepts and Tools

2.1 Robot 6-SPS

A 6-SPS robot (spherical- prismatic-spherical) is used in this work to demonstrate the method; where the base is connected to the platform [12,15].

This connection takes place through 6 serial kinematic chains that have a spherical joint attached to the base, a prismatic joint (actuated) and a spherical joint attached to the platform, where the end effector (tool or claw) should be fixed as shown in Fig. 1.

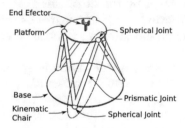

Fig. 1. A specific class of 6-SPS robot is the MSSM robot (Minimal Simplified Symmetric Manipulator) which has triangular base and platform.

2.2 Workspace

In order to implement this method it is necessary to know the parameters that limit the workspace. Limitation that restricts the robot to advance to some locations or directions. Singularities in direct and inverse kinematics are factors that limit the workspace for serial and parallel robots [19]. Collisions between the legs or legs and platform are factors that affect the workspace directly as well [8].

The workspace determination for a fixed end effector orientation is called position workspace [3, 10]. Whereas the workspace determination for a fixed end effector position is named orientation workspace [2, 3, 6]. Another workspace approach consists in partially join the orientation and position workspace, graphically disposing in a mixed representation [5].

2.3   Mapping

The workspace mapping is defined using indices indicating proximity to inoperative regions, i.e. regions outside the workspace. Then the indices for collisions, direct and inverse kinematics are analyzed. In order to present a unique graphical representation, it is necessary to unify the indices into a binary index, 0 or 1, for orientation where the platform is operative or inoperative, respectively. Aiming at representing workspace orientation, all the rotations are considered in the range from minimal angle ($Ang_{min} = -180°$) to maximal angle ($Ang_{max} = 180°$) with small step ($Ang_{step} = 2°$). In mixed workspace representation, it is necessary to consider linear displacement range from a minimal linear position (Lin_{min}) to a maximal linear position (Lin_{max}) with a linear step (Lin_{step}), that depends on robot configuration or type (see Appendix for details).

2.3.1   Inverse Kinematic Index

Inverse kinematic index (IKI) is considered when inverse kinematic singularities occur. Larger regions of robot orientation workspace are restricted if compared to regions restricted by the direct kinematic singularities. There are guidelines where the platform loses some degree of freedom due to the actuator stroke limits (maximum and minimum) when changing the robot platform orientation.

In order to describe correctly a rigid body in space it is necessary to know its position and orientation [17]. Robot orientation may be described by the angles Roll, Pitch and Yaw (ϕ, θ, ψ), that represents successive rotations around the axis X, Y and Z [17]. Therefore, robot rotation is the resultant rotation on axes. The robot position may be described by the vector $\overrightarrow{p}$ (see Fig. 2).

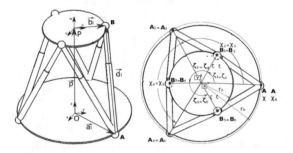

Fig. 2. Robot 6-S<u>P</u>S - Inverse kinematic

In order to measure the prismatic joint lengths (actuated joints), a vector $\overrightarrow{d_i}$ is considered, joining $\overrightarrow{a_i}$ to $\overrightarrow{b_i}$, for $i = 1...6$ (see Fig. 2). The points A_i represent spherical joints around the robot base, which are calculated considering the base angles χ_i and the base radius r_b, for $i = 1...6$. Likewise, B_i is calculated considering platform radius r_p and platform angles ζ_i.

Inverse singularity determination regards the instantaneous value of prismatic joint action $d_i = \vec{d_i}/\left\|\vec{d_i}\right\|$ and $\vec{d_i}$ may be obtained by

$$\vec{d_i} = \begin{bmatrix} p_x + (\cos(\phi)\sin(\psi)\sin(\theta) - \sin(\phi)\cos(\psi))\sin(\zeta_i)\,r_p + \cos(\phi)\cos(\theta)\cos(\zeta_i)\,r_p - r_b\cos(\chi_i) \\ p_y + (\sin(\phi)\sin(\psi)\sin(\theta) + \cos(\phi)\cos(\psi))\sin(\zeta_i)\,r_p + \sin(\phi)\cos(\theta)\cos(\zeta_i)\,r_p - r_b\sin(\chi_i) \\ p_z + \sin(\psi)\cos(\theta)\sin(\zeta_i)\,r_p - \sin(\theta)\cos(\zeta_i)\,r_p \end{bmatrix}. \quad (1)$$

Resulting in: $L_{min} \geq d_i \geq L_{max}$, limiting the joint to the value range for a certain orientation. The minimum value of prismatic joint action is L_{min} and maximum value of prismatic joint action is L_{max}. Values of d_i outside of the range will be considered inverse singularity points [3].

2.3.2 Direct Kinematic Index

The Direct Kinematic Index (DKI) indicates proximity to direct kinematic singularities, i.e, when the robot is in an instantaneous pose it gains some degree of freedom. Such indexes may be calculated based on virtual power coefficients [7,18]. Additionally, instantaneous robot behavior near singularity is described by a particular methodology [20].

The singularity work-based measurement is used to minimize the problem where the objective function is subjected to a constrained optimization problem. The constrained optimization can be solved using an unrestricted optimization problem. The transformation is obtained by a specific Lagrange formulation solving the optimization problem then resulting in unrestricted situation. This problem can be expressed in eigenvalues generalized problem. When the eigenvalues become zero, $\sqrt{\lambda_{min}} = 0$, they represent a direct kinematic singularity. This methodology is clearly described in different works [7,11,18].

Theoretically, the index value should be zero, but due to the gap in the joints, the index value for singularity is slightly higher [13]. In this case, the threshold value is established experimentally as described by [11], where the value is $\sqrt{\lambda_{min}} \leq 0.028$. In the present work no experimental analysis were performed, then the value of $\sqrt{\lambda_{min}} \leq 0.03$ is used.

2.3.3 Collision Proximity Index

In order to calculate Collision Proximity Index (CI) it is necessary to calculate collision between bodies. In this case, the robot legs are considered to have a radius r_a, attached to a collision tolerance between the legs, arriving at a linear collision tolerance distance TC_{lin} and an angular collision tolerance distance TC_{ang}. Thus the distance between the legs should not be less or equal to TC_{lin} and TC_{ang}. These distances measurements between the two legs are considered to be the starting values from the works of [8,15].

3   Maximization Based on Initial Platform Orientation

The proposed graphical optimization method maximizes the platform rotation. From the orientation workspace representative sections, the rotation maximization is obtained by calculating the convenient initial orientation (IO), represented by the initial orientation point (IOP) in graphical representation, see Fig. 3. After determining the IOP, the tool is compensated if necessary.

Considering the orientation workspace, graphical optimization of the platform rotation with different criteria is possible. In this work, two criteria are presented concerning the maximum robot platform rotation. In the first criterion, the maximization of the platform rotation around an axis situated in a plane is desired. It is represented by the line of maximum rotation axis (LMR) in the graphical representation, see Fig. 3. In the second criterion, it is desired to maximize the minimum platform rotation of a parallel robot around all the axis contained in a plane. Represented by the circle radius of minimum maximum rotations (CRMM), see Fig. 3.

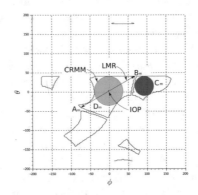

Fig. 3. Slice representation of orientation workspace (Rotation in ϕ, θ and $\psi = 60°$)

Figure 3 represents the workspace section (planes of rotations for two variables) considering variations in Roll and Pitch with fixed Yaw in $\psi = 60°$. LMR is represented by straight line $\overline{A_m B_m}$, which demonstrates the largest angular magnitude. The D_m circle (green) describes the CRMM, which is the region where the maximum and minimum rotations are obtained in that plan. The C_m circle region (blue) represents an example of the existence of more than one operative sub-space in the plane.

3.1   Tool Reorientation

When IOP is determined and the platform is guided in the convenient orientation ($\phi = \theta = \psi = 0$), the tool orientation must be compensated, since it is fixed to the platform. Thus, after platform reorientation and tool compensation, the

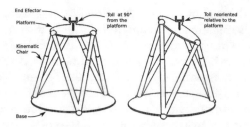

Fig. 4. Robot tool: original orientation and compensation

tool posture must remain the same in the global coordinate system; however, the platform posture may change (see Fig. 4).

After offsetting the tool orientation, the robot is ready to perform the task but, with largest rotation. If the IOP is $\phi = \theta = \psi = 0$, the compensation of the tool orientation is unnecessary.

Accordingly, the method may be detailed by following seven steps:

1. *Map the workspace.* The platform is considered to have the IO ($\phi = \theta = \psi = 0°$), initial position ($p_x, p_y, p_z$) and robot constants.
2. *Workspace Slices Representation (rotation planes).* Separate the calculated orientation workspace into slices. These sections represent Roll, Pitch or Yaw rotation and position displacement in p_x, p_y, p_z. For a 3D representation, three information to vary and three information to fix are needed.
3. *Choose the section that has the highest LMR/CRMM included within an operative region.* Considering interval and step, 181 orientation workspace sections are available. One of these sections contains the highest LMR/CRMM, which represents the maximum rotation. The determination of the line/circle is performed visually in each section.
4. *Determine the IOP, which indicates the initial platform orientation. Two cases are considered:*
   a. LMR: If the rotation is symmetric, IOP is situated at the LMR center (conventioned F_m). If the rotation is asymmetric, IOP is situated at one end of the line, and should be at the points A_m or B_m. Considering the orientation workspace section that has the highest LMR is determined, the convenient initial platform orientation represented by the IOP must be calculated.
   b. CRMM: Considering the workspace section that has the largest CRMM is determined, the convenient platform initial orientation represented by the IOP must be calculated, that is represented by the center of CRMM.
5. *Determine the maximum rotation. Two cases are considered:*
   a. LMR, by measuring the length of:
      i. if the rotation is performed in one axis direction only, IOP is represented by the LMR.
      ii. if the rotation is performed in both axis directions, this should be represented by LMR/2.

  b. CRMM: In determining the CRMM one can determine its radius, which
  represents the maximum minimum turn in all directions of rotation.

6. *Platform pose according to the initial orientation, indicated by the IOP.* Conveniently when starting from IOP, the maximum platform rotation is maximized. Thus, the change to the convenient initial orientation allows a greater rotation than those obtained in the conventional initial orientation.

7. *Compensate tool orientation.* The tool compensation angles represents the difference between the initial orientation and the convenient initial orientation (see Fig. 4).

4 Method Application

The presented method is applied to the parallel robot MSSM [15]. The robot dimensions are: base radius $r_b = 0.7$ m, platform radius $r_p = 0.3$ m, linear collision tolerance $TC_{lin} = 0.05$ m, angular collision tolerance $TC_{ang} = 8°$, actuator lower limit $L_{min} = 0.4$ m and upper limit $L_{max} = 1.3$ m. Emphasizing that the platform is modeled in a rigid structure of cylindrical rays that joint its origin with the connection points to the spherical joints.

4.1 Task 1

It is desired to observe the passage of a comet using a telescope. In order to record this event, the determination of the maximum rotation on a rotation axis in XY($\phi\theta$) plane is necessary for the MSSM robot. The platform position is set to $p_x = 0$ m, $p_y = 0$ m and $p_z = 0.8$ m.

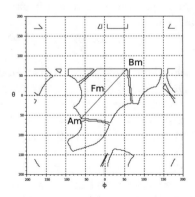

Fig. 5. Orientation workspace slice that maximizes line of maximum rotation axis

The points where IOP may be situated (A_m, B_m e F_m) are represented in Fig. 5. The largest asymmetric rotation obtained is 188° for Yaw of 30°. This rotation has been confirmed using MSC ADAMS. The reorientation depends on where one intends to start the rotation: (a) case IOP $= A_m$, the initial orientation is $\phi = -66°, \theta = -65°Y$; (b) case IOP $= B_m$, the initial orientation is $\phi = 61°, \theta = 73°$. Otherwise, the largest symmetric rotation is 94° for Yaw of 30°. The reorientation is determined by IOP $= F_m = (\phi = 2°, \theta = 3°)$. Before maximization, the maximum rotation was 120°, so there is a gain of 68° in the asymmetric rotation and of 34° in the symmetric rotation by using the proposed optimization method.

4.2   Task 2

For a highly complex part manufactured in series by polymer injection, superficial tests to gauge the part measurements are carried out. For this measurement a tool is fixed to the robot platform, and optimization of rotation in all directions in the plane of rotations around axis X and Y (ϕ, θ), such as the coordinate p_z that maximizes the rotation are performed. The platform is considered in initial orientation $\phi = \theta = \psi = 0$ and initial position $p_x = 0$ m, $p_y = 0$ m and $p_z = 0.4$ m.

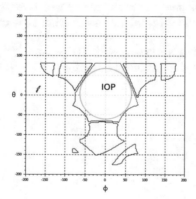

Fig. 6. Orientation workspace slice that maximizes radius of minimum maximum rotations

The point where IOP is determined by the CRMM center is represented in Fig. 6. The largest circumscribed circle has a radius of 68°, representing the largest symmetric turn in the plane for a $p_z = 0.644$ m, where the reorientation is $\phi = 0°$ and $\theta = 4°$. Initially, the maximum rotation was 66°, and then with maximization there is a gain of 2° in all directions.

5   Conclusion

Robot workspace representations are the results of a unified binary "viability" index. The unified index is calculated from other robot indexes (singularity and collision). It is observed that the order in which these indexes are calculated influence the computational method cost, reducing the computing time by up to two thirds of the time.

In order to demonstrate that the initial orientation may maximize the desired parameters for a certain task, two tasks were performed with the MSSM robot, demonstrating the performance for each task. Task 1 determines the largest possible rotation in an axis in XY ($\phi\theta$) plane. In task 2, the largest possible rotation on all ϕ, θ plane axis was determined.

This method allows maximizing the platform rotation of a parallel robot without structural modifications, considering initial platform and tool reorientation only. In addition, it was observed that there are other viable sub-spaces for the module that departs from the conventional orientation ($\phi = \theta = \psi = 0°$), which in some occasions may be a solution for greater turning in one direction. To move from one viable sub-space to another it is necessary to disassemble and reassemble the robot, transposing barriers as the collision between bodies.

Perform the computational automation of the CRMM, LMR and IOP search process for the plans within the mapping algorithm is necessary. The next step consists in three-dimensional mapping to see which largest sphere will fit within Orientation Workspace and its initial orientation that maximizes the spins.

Acknowledgements. This work was partially supported by 'Fundação Coordenação de Aperfeiçoamento de Pessoal de Nível Superior (CAPES), PGPTA 59/2014 AUXPE 3686/2014.

Appendix

```
 1: function Mapping( r_b, r_a, χ_{1..6}, ζ_{1..6}, L_min, L_max, TC_lin, TC_ang, ...)
        (...Lin_min, Lin_max, Lin_step, Ang_min, Ang_max, Ang_step)
 2:     for φ = Ang_min : Ang_step : Ang_max do              . Choose three of six
 3:         for θ = Ang_min : Ang_step : Ang_max do          . Choose three of six
 4:             for ψ = Ang_min : Ang_step : Ang_max do      . Choose three of six
 5:                 for p_x = Lin_min : Lin_step : Lin_max do  . Choose three of six
 6:                     for p_y = Lin_min : Lin_step : Lin_max do  . Choose three of six
 7:                         for p_z = Lin_min : Lin_step : Lin_max do  . Choose three of six
 8:                             if DistanceBetweemBodies ≤ TC_lin then
 9:                                 CI = 0;                            . Operative
10:                             else
11:                                 if DistanceBetweemBodies > TC_lin then
12:                                     CI = 1;                        . Inoperative
13:                                     Return 1;
14:                                 end if
15:                             end if
16:                             if AngleBetweemBodies ≤ TC_ang then
17:                                 CI = 0;                            . Operative
18:                             else
19:                                 if AngleBetweemBodies > TC_ang then
20:                                     CI = 1;                        . Inoperative
21:                                     Return 1;
22:                                 end if
23:                             end if
24:                             if L_min ≥ di; di ≤ L_max then
25:                                 IKI = 0;                           . Operative
26:                             else
27:                                 if L_min < di then;
28:                                     IKI = 1;                       . Inoperative
29:                                     Return 1;
30:                                 end if
31:                             else
32:                                 if di > L_max then
33:                                     IKI = 1;                       . Inoperative
34:                                     Return 1;
35:                                 end if
36:                             end if
37:                             if √λ_min > 0.03 then
38:                                 DKI = 0;                           . Operative
39:                             else
40:                                 if √λ_min ≤ 0.03 then
41:                                     DKI = 1;                       . Inoperative
42:                                     Return 1;
43:                                 end if
44:                             end if
45:                         end for    . When index is 1 the robot pose is inoperative, else every
                                         indices are 0, i.e, is operative pose. That occurs for every
46:                     end for                pose into the ranges..
47:                 end for
48:             end for
49:         end for
50:     end for
51: end function
```

References

1. Arora, J.: Introduction to Optimum Design. Academic Press, Cambridge (2004)

2. Arrouk, K., Bouzgarrou, B.C., Gogu, G.: On the workspace representation and determination of spherical parallel robotic manipulators. In: New Trends in Mechanism and Machine Science, pp. 131–139. Springer, Heidelberg (2017)

3. Au, W., Chung, H., Chen, C.: Path planning and assembly mode-changes of 6-DOF stewart-gough-type parallel manipulators. Mech. Mach. Theory **106**, 30–49 (2016)

4. Bhatti, M.A.: Practical Optimization Methods: With Mathematica® Applications. Springer, Heidelberg (2012)

5. Bohigas, O., Manubens, M., Ros, L.: A linear relaxation method for computing workspace slices of the stewart platform. J. Mech. Robot. **5**(1), 011005 (2013)

6. Bonev, I.A., Ryu, J.: A new approach to orientation workspace analysis of 6-dof parallel manipulators. Mech. Mach. Theory **36**(1), 15–28 (2001)

7. Brinker, J., Corves, B., Takeda, Y.: Kinematic performance evaluation of high-speed delta parallel robots based on motion/force transmission indices. Mech. Mach. Theory **125**, 111–125 (2018)

8. Brisan, C., Csiszar, A.: Computation and analysis of the workspace of a reconfigurable parallel robotic system. Mech. Mach. Theory **46**(11), 1647–1668 (2011)

9. Cammarata, A.: Optimized design of a large-workspace 2-DOF parallel robot for solar tracking systems. Mech. Mach. Theory **83**, 175–186 (2015)

10. Gokul Narasimhan, S., Shrivatsan, R., Venkatasubramanian, K., Dash, A.K.: Determination of constant orientation workspace of a stewart platform by geometrical method. In: Applied Mechanics and Materials, vol. 813, pp. 997–1001. Trans Tech Publications (2015)

11. Hesselbach, J., Bier, C., Campos, A., Lowe, H.: Direct kinematic singularity detection of a hexa parallel robot. In: Proceedings of the 2005 IEEE International Conference on Robotics and Automation, pp. 3238–3243. IEEE (2005)

12. Jiang, Q., Gosselin, C.M.: Determination of the maximal singularity-free orientation workspace for the gough-stewart platform. Mech. Mach. Theory **44**(6), 1281–1293 (2009)

13. Last, P., Budde, C., Bier, C., Hesselbach, J.: Hexa-parallel-structure calibration by means of angular passive joint sensors. In: 2005 IEEE International Conference on Mechatronics and Automation, vol. 3, pp. 1300–1305. IEEE (2005)

14. Li, W., Angeles, J.: The design for isotropy of a class of six-dof parallel-kinematics machines. Mech. Mach. Theory **126**, 34–48 (2018)

15. Merlet, J.: Parallel Robots. Kluwer Academic Publisher, Boston (2000)

16. Portman, V., Chapsky, V., Shneor, Y.: Workspace of parallel kinematics machines with minimum stiffness limits: collinear stiffness value based approach. Mech. Mach. Theory **49**, 67–86 (2012)

17. Siciliano, B., Sciavicco, L., Villani, L., Oriolo, G.: Robotics: Modelling, Planning and Control. Advanced Textbooks in Control and Signal Processing, vol. 26, p. 29. Springer, London (2009)

18. Voglewede, P.A.: Measuring closeness to singularities of parallel manipulators with application to the design of redundant actuation (2004)

19. Wang, X., Zhang, D., Zhao, C., Zhang, H., Yan, H.: Singularity analysis and treatment for a 7R 6-DOF painting robot with non-spherical wrist. Mech. Mach. Theory **126**, 92–107 (2018)

20. Wolf, A., Shoham, M.: Investigation of parallel manipulators using linear complex approximation. ASME J. Mech. Des. **125**(3), 564–572 (2003)

CNut Gathering Robot. Design, Implementation and Mathematical Characterization

Rafaela Villalpando-Hernandez[✉], Cesar Vargas-Rosales,
Rene Diaz-M., Lizeth Espinoza, and Alberto Martínez

Tecnologico de Monterrey, Monterrey, Mexico
{rafaela.villalpando, cvargas, renejdm}@tec.mx,
izet2097@gmail.com, jalbertomtzcarrillo@gmail.com

Abstract. In recent years, robotics has become an important knowledge field to support development in different areas, such as agriculture. The current methods of walnut gathering are either expensive or cumbersome. In this paper, the design, kinematics analysis and the implementation of a low cost, low complexity walnut gathering robot are presented. The proposed robot in this paper, was build based on a mechanical structure capable of doing the required movements and actions to move throughout rough terrains and collect the walnuts automatically at a very low cost.

1 Introduction

Robotics applied to agriculture is an important for the development of a nation, as it can help to increase production and reduce risk, human errors and accidents. "La Laguna" region in the northern Mexico is the second place in the nation in walnut production. The current process for collecting walnuts is long and expensive, first a machine shakes the trees, so the walnuts fall on the floor, then with the help of a mesh the walnuts are collected, [1]. Another method is by using big pieces of mesh lying on the ground, so the farmers can pick up the walnuts that fall there. These methods can result costly, and inefficient, as a high percentage of the walnuts is left on the ground. Several robot's proposals for walnut re-collection have been proposed. Authors in [2], use a mesh above a robot combined with a tree shaking system, so the nuts fall on the mesh. This robot results costly, due to the large motors necessary to implement the tree-shaking system. In [2], a company presents an attachment to a tractor, this goes into the hand of the tractor and the walnuts fall onto it.

In this paper, the design and construction of a low-cost, low complexity and light option robot is proposed. Its objective is to gather the walnuts in a way that any farmer could use it. The method chosen for the recollection of the walnuts is the hamster ball. The movement mechanism was designed so that the robot could adapt to uneven terrains, therefore a wheel base or a continuous track mechanism was chosen, the construction materials were selected to be light and of low-cost. The mechanical design was done so that the robot could be robust, and easy to maintain.

A. Martínez et al. (Eds.): LACAR 2019, LNNS 112, pp. 51–63, 2020.
https://doi.org/10.1007/978-3-030-40309-6_6

2   Mechanical Design

There are several issues to consider in the robot mechanical design. In this section, we present the characteristics of the selected locomotion and nut collection systems as well as the mechanical robot design and construction.

2.1   Locomotion Mechanism

The movement method was designed according to the robot's application and considering the terrain in which the robot would operate. There have traditionally been two robotic locomotion methods, wheels and tank tracks, [3]. These methods are effective in several surfaces. While a tank track robot is far more flexible and can cross a variety of terrains, wheeled robots are efficient for crossing large, flat landscapes, but they struggle when their path contains obstacles. Advantages and disadvantages of locomotion systems are described in Table 1.

Table 1. Characteristics of the locomotion systems.

Systems	Costs	Velocity	Obstacles	Traction	Simplicity
Wheels	Low cost	Low speed	Not very efficient	Low traction	Less moving parts
Tank tracks	High cost	High Speed	Very efficient	High traction	Difficult mechanisms

According to the characteristics in Table 1 and since the robot will be operating in rough and uneven surfaces, tank tracks were selected as the locomotion system.

2.2   Collection Mechanism

Different types of mechanisms were studied for the walnut gathering system. In [4], authors present several types of gatherers as the nut harvester by carrot design. The strong point in this one is its versatility since it can pick from small nuts like acorns to big ones like hazelnuts, however its downside is the fact that it is predisposed to rusting, so it would not last in the kind of environment envisioned for the robot. In [5], a cyclone rake system was used, that consist of a cylinder with "hair" that allows the nuts to be stuck between the columns in the cylinder. This mechanism can result expensive due to the cost of the large motor needed to move the cylinder.

The "hamster wheel" mechanism results in a low cost and lightweight option for gathering. This mechanism picks the nuts by exerting pressure between the metal, the walnut and the floor, this force pushes away the metal rods, so it can allow the nuts inside the structure.

2.3   Robot's Design and Construction

The final design of the robot could be described as a catapult with wheels, see Fig. 1. The structure was designed to provide stability and the necessary weight to support the wheel and to apply sufficient force to gather the nuts.

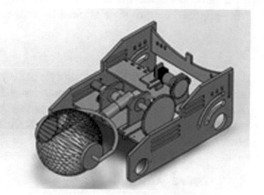

Fig. 1. Robot-s CAD assembly.

To be able to move the nut gatherer, a mechanism of gears coupled with a reduction was used, so the motor could provide a torque of 1.57 N-m to lift the nut gatherer. A counter weight was also added for extra support to the motor.

The construction of the robot chassis was made using 6 mm wide medium-density fibre (MDF) board. To keep the robots low-cost, the tracks were manually built recycling a bike chain as it is seen in Fig. 2. The chains move thanks to four sprockets inside of them just like in a bike mechanism. The sprockets were incorporated (with couplings made of polylactic acid (PLA) in a 3D printer) to the motors to increase the torque.

Fig. 2. Nut gathering robot final assembly.

3   Electronic Design

The robot moves with two primary sprockets located at the back, made up by two identical high power gearmotors with a 500:1 reduction. The rotation of the end-effector to nuts collector is powered by a gearmotor with a 30:1 reduction. Due the high torque needed to accomplish both the driving and the nut gathering tasks, low and high-power electronics with optimized control signals must be applied.

3.1   Power Supply

To use the robot at full capacity for one hour, an electrolyte battery of 6 V and 7Ah was selected based on the power characteristics of each motor, see Table 2 for specifications. The development platform used for the project was the Arduino Mega board, which can be supplied by 7–12 VDC, so an ordinary 9 V alkaline battery was used to energize this low power circuit.

Table 2. Gearmotors specifications.

Function	Motor description	Velocity	Reduction	Max voltage	Max current
Back sprockets drivers	DC Gearmotor	12 rpm	500:1	6 V	2.5 A
Walnut collector	DC Gearmotor	30 rpm	300:1	12 V	3 A

3.2   H Bridge DC Motor Drive

To enable the robot to go forward or backward, and make complete left or right turns, the motors should be able to speed counterclockwise and clockwise. To accomplish this, it was necessary to implement an H bridge circuit, [6], for the control of each motor. Transistors (TIP41C), were used for current amplification. These are NPN bipolar transistors that can stand up to 10 A and 100 V.

3.3   Speed Control

Two pulse wide modulated (PWM) signals activate the transistors, to control the speed and direction of motors, [7]. These signals are generated through the analog output function of the Arduino. Speed control is crucial for the walnut collector, because it is necessary to increase the inertia of this tool when it goes completely down. Also, six PWM signals are needed, three for the clockwise direction of each motor and three for the counterclockwise rotation.

3.4   Steering Drive Relations

To turn left and right, a differential driving system was chosen, its geometric representation is shown in Fig. 3. To maintain the algebraic manipulation as simple as possible, the longitude V_R is approximated to 0, therefore distance R depends exclusively on longitude W. Also note, that V_L has a direct relation with the PWM duty cycle, assuming there are no losses and the maximum angular velocity is 12 rpm. Distance R is calculated as follows

$$R = \frac{W(V_L - V_R)}{2(V_L + V_R)} \tag{1}$$

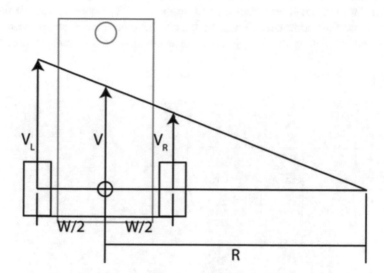

Fig. 3. Steering drive representation for a tank chassis.

3.5   Electronic Improvements

The first improvement to be made is to add encoders at least to the motors in the chassis driving. Even if the motors come exactly from the same supplier, they have variations in their velocity that makes impossible for the robot to move straight forward or backwards. A correction in velocity needs to be made, with the feedback of the encoder and a proportional (P) or proportional and integral (PI) controller, it is possible to correct the robot's trajectories.

4   Communications

Related to communications, it was a priority to establish a friendly interface between users and robot. An android application (Bluetooth serial controller) gives the chance to add and edit the appearance, name and message which will be transmitted using Bluetooth communication. Bluetooth used is the module HC-06, [8].

4.1   Bluetooth Serial Controller App Configuration

The final user interface had seven buttons: forward, backward, right, left, nut collector up, nut collector down and stop. Every time the user presses a button, a command configured with this app, is sent using Bluetooth. The Bluetooth device needs to be chosen in the smartphone configuration and then settled in the serial controller app.

The serial communication standard baud rate is 9600, [7], the serial port number one of the Arduino mega was used for this application. The code to control the robot consists on starting communications, analyzing the command received with a case structure and activating the PWM ports corresponding to the motors needed. See Fig. 4.

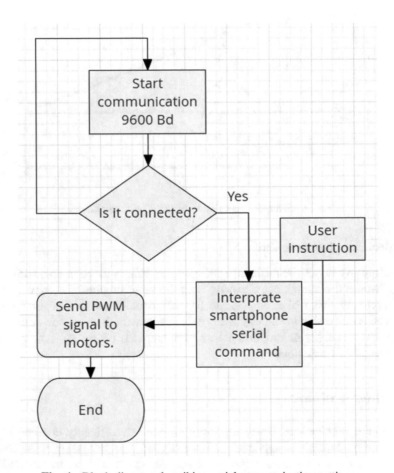

Fig. 4. Block diagram describing serial communication settings.

5 Robot Mathematical Characterization

To be able to control the robot's movements the following motion analysis were developed: forward kinematics, inverse kinematics and Jacobian calculations. All of them focused on the movement description of the nut collector. This with the objective of avoiding the human intervention for the robot operations.

5.1 Forward Kinematics

The robot has only two degrees of freedom and three motors, so only two frames are needed to perform the direct kinematics. Figure 5 shows the frames which were located using the Denavite-Hartenberg rules and variables, [7], coordinate rules. In Table 3, the parameters of the robot geometry (see Fig. 6) are presented, where Q is the angle between X_n and X_{n+1} relative to Z_n. A is the angle between Z_n and Z_{n+1} relative to X_n, (refer to front view shown in Fig. 6c), in case of frames 1 and 2 the angle between their Z axes is zero.

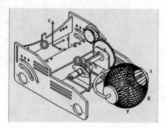

Fig. 5. Frames allocation in the robot, for direct and inverse kinematics.

Table 3. Frames description for forward kinematics

Frame	Q	A	R	D
1	Q_1	90°	L_1	0
2	Q_2	0	L_2	0

The parameter R describes the distance between the origin of each frame measured through the X_n axis. The last parameter, d refers to the distance between the origin of each frame measured through the Y_n axis, frames 0, 1 and 2 are at the same height and in the same center line for this analysis, so the value of d equals to 0. The distribution of the frames is under the assumption that both wheels move at opposite velocities, so the robot is performing rotation as shown in Fig. 7.

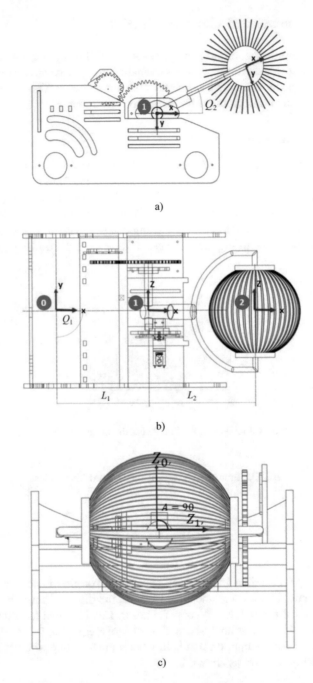

Fig. 6. Geometric references to compute forward kinematics. (a) Lateral view. (b) Top view. (c) Front view.

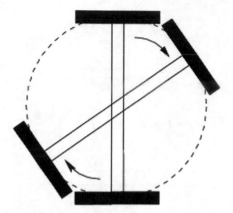

Fig. 7. Rotation. Wheels move at the same speed in opposite direction.

As stated before in Fig. 5, there are a total of three frames. Frame 0 translates and rotates to frame 1, then there is a translation to frame 2. The movement description of the robot is supported by its geometric data shown in Fig. 8, where distances between frames L_1, and L_2 are illustrated, also angles Q_1 and Q_2. The translation and rotation final matrix operation give the following expressions, please let us consider that $s_1 = \sin Q_1$, $c_2 = \cos Q_2$ and R_{0-2} refers to the translation and rotation matrix description from frame 0 to 2.

$$R_{0-2} = \begin{bmatrix} r_1 & r_4 & r_7 & r_9 \\ r_2 & r_5 & r_8 & r_{10} \\ r_3 & r_6 & \frac{\sqrt{2}}{2} & r_{11} \\ 0 & 0 & 0 & 1 \end{bmatrix} \tag{2}$$

Where r_k is an abbreviation of the following equations,

$$r_1 = c_1 c_2 - \frac{\sqrt{2} s_1 s_2}{2}$$

$$r_2 = -c_1 s_2 - \frac{\sqrt{2} s_1 s_2}{2}$$

$$r_3 = \frac{\sqrt{2} s_1}{2}$$

$$r_4 = L_1 c_1 + L_2 c_1 c_2 - \frac{\sqrt{2} L_2 s_1 s_2}{2}$$

$$r_5 = c_2 s_2 + \frac{\sqrt{2} c_1 s_2}{2}$$

$$r_6 = -s_1 s_2 + \frac{\sqrt{2} c_1 c_2}{2}$$

$$r_7 = \frac{-\sqrt{2} c_1}{2}$$

$$r_8 = L_1 s_1 + L_2 c_1 s_2 + \frac{\sqrt{2} L_2 c_1 s_2}{2}$$

$$r_9 = \frac{\sqrt{2} s_2}{2}$$

$$r_{10} = \frac{\sqrt{2} c_2}{2}$$

$$r_{11} = \frac{\sqrt{2} L_2 s_2}{2}$$

Recall that the transformation matrix $R_{i\text{-}j}$ of the Denavit-Hartenberg method represents the resumed mathematical operations to calculate de position of the final robot actuator P_X, P_Y when some construction variables (L_1, L_2, etc.) are stated. Therefore, r_k only represent intermediate algebraic operations.

5.2   Inverse Kinematics

This robot has two independent motion systems. In the inverse kinematics characterization, the P_X, P_Y, P_Z coordinates of the end effector with respect to frame 0 are given. The following is the study of the movement of the first motion system, the Nut Gathering showed in Figs. 8 and 9. Angle Q_2 can be expressed as follows

$$Q_2 = \arctan\left(\frac{P_2}{P_1}\right) \tag{3}$$

$$P_2 = P_Z \tag{4}$$

$$P_1 = \sqrt{L_2^2 - P_Z^2} \tag{5}$$

Note that L_2 is shown in Fig. 6(b). On the other hand, we have the second system where angle Q_1 can be computed using P_X and P_Y coordinates directly as follows

$$Q_1 = arctan\left(\frac{P_y}{P_x}\right) \tag{6}$$

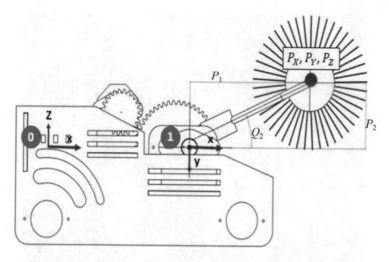

Fig. 8. Lateral view and references to compute inverse kinematics.

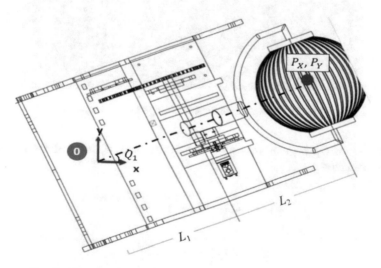

Fig. 9. Top view and references to find Q_1 in inverse kinematics.

5.3 Jacobian for Velocity Analysis

A Jacobian matrix is equivalent to the derivative of a vector valued function, [7]. Using Table 3, an algorithm was programed to find the velocity description as a matrix equation, which is

$$j(\omega_x, \omega_y, \omega_z) = \begin{bmatrix} M_1 & M_3 & 0 \\ M_2 & M_4 & M_5 \end{bmatrix} \tag{7}$$

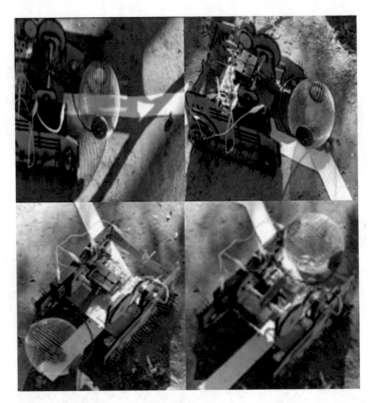

Fig. 10. Walnut gathering robot operational sequence

Where ω_n is the angular velocity for each axis and M_x is an abbreviation of the following equations:

$$M_1 = -L_1 s_1 - L_2 c_2 s_1 - \frac{\sqrt{2} L_2 c_2 s_1}{2}$$

$$M_2 = -L_2 \left(2 c_1 s_2 + \frac{\sqrt{2} c_1 s_1}{2} \right)$$

$$M_3 = -L_1 c_1 - L_2 c_1 c_2 - \frac{\sqrt{2} L_2 s_1 s_2}{2}$$

$$M_4 = L_2 s_1 s_2 - \frac{\sqrt{2} L_2 c_1 c_2}{2}$$

$$M_5 = \frac{\sqrt{2} L_2 c_2}{2}$$

Matrix in Eq. (7) represents the velocity of the end effector using frame 0 as reference. Recall that, $s_1 = sinsinQ_1$, and $c_2 = coscosQ_2$.

Finally, in Fig. 10, we present evidence of nut gathering robot functionality. In this figure, a sequence is shown from the beginning of a programed trajectory, then the robot is shown moving from a curve and the nut collection at the back side of the robot is shown.

The robot gathers nuts easily and with precision, but it lacks accuracy following trajectories due the low quality of the motors installed on the tracks. Currently the robot can only be controlled manually, but a velocity analysis was described before, and it may be used to get feedback from motors with encoders to develop an autonomous mode.

6 Conclusions

In this paper, we present the design and implementation and kinematic analysis of a low cost, power efficient, and robust nut gathering robot. Through this work a proof of concept of a very low cost (less than $200.00USD) nut gathering robot is presented with evidence of functionality.

For future work, we propose an improvement to the mechanical design, installing higher torque motors. Also, the PLA parts and the chassis can be replaced by aluminum parts, so the weight could be reduced, and the construction robustness increased.

References

1. Vázquez, N.C.R.: Retos y oportunidades para el aprovechamiento de la nuez pecanera en México. CIATEJ 112 (2016)
2. VIMAR EQUIPOS S.L. (2018). Vimar. https://vimarequipos.com/
3. Into robotics, 11 November 2013. https://www.intorobotics.com/wheels-vs-continuous-tracks-advantages-disadvantages/
4. Nut Gatherers (2018). https://nutgatherers.com/
5. Woodland Power Products (2017). Cyclone Rake. https://www.cyclonerake.com/cyclone-nut-rake-for-pecans/
6. Sedra, A.S., Smith, K.: Microelectronic Circuits, 7th edn. Oxford University Press, Oxford (2014)
7. Siciliano, B.: Robot foundations: kinematics. In: Springer Handbook of Robotics, pp. 9–33. Springer, Heidelberg (2008)
8. Grimmett, R.: Going truly mobile – the remote control of your robot. In: Arduino Robotics Projects, pp. 137–156. Packr Publishing Ltd, Birmingham (2014)
9. Craig, J.: Introduction to Robotics, pp. 135–140. Pearson Education, Upper Saddle River (2005)
10. Mohan, N.: Power Electronics, pp. 368–380. Wiley, New Delhi. MacKerrow (1998)
11. Rumsey, F., Watkinson, J.: Digital Interfaces Handbook, pp. 88–115. Elsevier/Focal Press, Amsterdam (2004)

Static Force Analysis of a Variable Geometry Legged Wheel

Héctor Moreno[1]([✉]), Oliver Zendejo[1], Isela Carrera[1], José Baca[2],
and Ismael Calderón[1]

[1] Universidad Autónoma de Coahuila, Barranquilla S/N, Guadalupe,
25750 Monclova, Coah., Mexico
h_moreno@uadec.edu.mx
[2] Texas A&M University-Corpus Christi, 6300 Ocean Dr, Corpus Christi, TX, USA

Abstract. This paper presents the static force analysis of a RSRR HeIse
wheel. This device consist of a 2 DOF mechanism that can transform a
circular wheel into a hybrid wheel with multiple limbs. The presented
analysis determines the forces acting on the joints of the mechanism and
the required forces at the actuators to achieve static equilibrium given
an external force applied at the traction link of the mechanism.

1 Introduction

Wheels and legs are two of the most popular methods of locomotion in mobile
robotics. Wheeled locomotion is the most efficient way to transit on plain terrains. On the other hand, the main advantage of legged locomotion systems is
their maneuverability in unstructured environments. Recently, some researchers
have proposed designs of hybrid locomotion systems based on wheels and legs.
Three categories have been proposed. The first one presents robots with legs
and wheels mounted on the vehicle's chassis [4,8]. In this case, when the robot
moves using the wheels the legs are retracted. The legs work when wheels are
not able to cross an obstacle. The second category corresponds to those robots
with wheeled legs. This locomotion system consist of a series of limbs with one
wheel attached at the end of each link [2,9]. There are also some designs that
include various wheels at different points of the legs. The last category consist of vehicles with legged wheels. These wheels consist of a symmetrical body
with multiple limbs, that is rotated by an actuator in the chassis of the vehicle
[3]. Some designs permit to control the extension or configuration of the limbs
[1,7,10].

In this paper we present the static force analysis of a RSRR HeIse wheel.
This mechanism can transform a circular wheel into a hybrid wheel with multiple
limbs. The kinematic chain of the mechanism allows the extension of the limbs
and the rotation of the wheel be independently controlled by two servomotors.
An interesting feature of the mechanism is that actuators are placed on the
chassis of the vehicle and not inside of the wheels, in this way the volume and
mass of the wheel is reduced. Additional features are described in [6]. HeIse

A. Martínez et al. (Eds.): LACAR 2019, LNNS 112, pp. 64–71, 2020.
https://doi.org/10.1007/978-3-030-40309-6_7

wheels were proposed by H. Moreno and I. Carrera and description of the whole family of mechanisms is presented in [5].

2   RSRR HeIse Wheel

Figure 1 shows a 3RSRR HeIse wheel. The mechanism of this device employs two actuators to control the configuration of the limbs. One actuator extends and flexes the limbs, and the second actuator rotates the hybrid wheel. As shown in Fig. 1, the mechanism consists of a sliding shaft, a rim, and three deployable limbs. The sliding shaft is connected to the triangular rim trough a prismatic joint. Also, the sliding shaft is connected to the rod by a rotational joint orthogonal to the axis of the wheel. The other end of the rod is connected by a spherical joint to the traction link. The traction link is connected by a rotational joint to the proximal link. Finally, the proximal link is connected to the rim by means of a rotational joint.

Before the static force analysis is presented, the forward kinematic analysis (necessary for the simulation in Sect. 3) is briefly introduced. The analysis consists in determining the position and orientation of the links of the mechanism given the values of the vector of joint variables $\mathbf{q} = \begin{bmatrix} q_1 & q_2 \end{bmatrix}^T$. To analyse this three-dimensional mechanism, it is noted that it can be separated into two parts, and each of these parts can be analyzed in one plane. Two reference frames are defined to do this. The reference frame Σ_1 is fixed to the robot chassis, and its z_1 axis is parallel to the wheel axis. On the other hand, the reference frame Σ_2 is attached to rim of the mechanism an its z_2 axis is parallel to z_1.

For the first part of the analysis consider Fig. 2. From this figure it is possible to obtain the following sum of vectors:

$$\mathbf{r}_1 + \mathbf{r}_2 + \mathbf{r}_3 = \mathbf{r}_4 \tag{1}$$

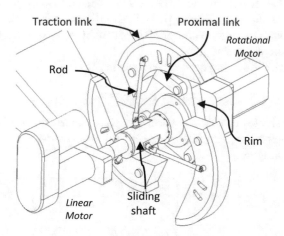

Fig. 1. 3RSRR HeIse wheel

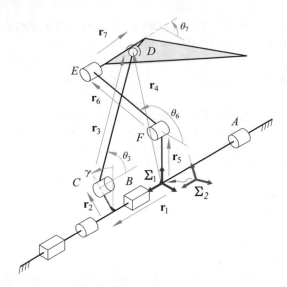

Fig. 2. Kinematic scheme of one leg of a 3RSRR wheel

The unknown variables in this equation are the orientation of vector $\mathbf{r}_3$ and the magnitude of $\mathbf{r}_4$, that are denoted by θ_3 and l_4, respectively. Those variables are given by the following expressions:

$$\theta_3 = \tan^{-1}\left(\pm\frac{\sqrt{l_3^2 - l_1^2}}{l_1}\right) \tag{2}$$

and

$$l_4 = l_2 \pm \sqrt{l_3^2 - l_1^2} \tag{3}$$

where l_2 is the distance from the wheel axis to the rotational joint C ($l_2 = |\mathbf{r}_2|$), l_3 is the length of the rod ($l_3 = |\mathbf{r}_3|$), l_4 is the distance from the end of the rod to the wheel axis. The magnitude of l_1 is given by $l_1 = l_p - q_p$, where l_p is the dead length of the prismatic joint.

From Fig. 2 the following sum of vectors is obtained:

$$\mathbf{r}_5 + \mathbf{r}_6 + \mathbf{r}_7 = \mathbf{r}_4 \tag{4}$$

the unknown variables in this equation are the orientation of vectors $\mathbf{r}_6$ and $\mathbf{r}_7$ in the $x_1 - y_1$ plane, i.e., the angles θ_6 and θ_7, respectively. The solution of the previous equation gives:

$$\theta_6 = 2\arctan\left(\frac{U_2 \pm \sqrt{U_1^2 + U_2^2 - K_1^2}}{K_1 + U_1}\right) \tag{5}$$

$$\theta_7 = 2\arctan\left(\frac{U_2 \pm \sqrt{U_1^2 + U_2^2 - K_2^2}}{K_2 + U_1}\right) \tag{6}$$

where $U_1 = l_4 \cos(\frac{\pi}{2} + q_r + \gamma) - l_5 \cos(\frac{\pi}{2} + q_r)$, $U_2 = l_4 \sin(\frac{\pi}{2} + q_r + \gamma) - l_5 \sin(\frac{\pi}{2} + q_r)$, $K_1 = (U_1^2 + U_2^2 + l_6^2 - l_7^2)/2l_6$ and $K_2 = (U_1^2 + U_2^2 + l_7^2 - l_6^2)/2l_7$. γ is the angle between $\mathbf{r}_5$ and $\mathbf{r}_2$, l_5 is the distance from the wheel axis to the rotational joint F ($l_5 = |\mathbf{r}_5|$), l_6 is the length of the proximal link ($l_6 = |\mathbf{r}_6|$), and l_7 is the distance from the rotational joint E to the universal joint D ($l_7 = |\mathbf{r}_7|$).

3   Static Force Analysis

This section presents the static force analysis of a RSRR HeIse wheel. The analysis determines the forces acting on the joints of the mechanism. The analytical expressions derived in this section can be used for design purposes. Even though the wheel is a mobile mechanism, if it moves at relatively low speeds, the static force analysis can be useful since inertial forces would be negligible compared with the external forces applied to the wheel and the links weights.

The unknowns in this analysis are the required torque in the rotational actuator to achieve the static equilibrium (τ_r), the force required in the linear actuator (τ_p), and the internal forces at the mechanism joints. In this study, the static equilibrium equations are obtained for each link. Figure 3 shows free-body diagrams of the links of one limb of the wheel.

The static equilibrium equations for the traction link are as follows:

$$\mathbf{f}_{d7} + \mathbf{f}_{e7} + \mathbf{F}_t + m_T\mathbf{g} = 0 \tag{7}$$

$$-\mathbf{r}_7 \times \mathbf{f}_{e7} + (\mathbf{r}_t - \mathbf{r}_7) \times \mathbf{F}_t + (\mathbf{r}_T - \mathbf{r}_7) \times m_T\mathbf{g} + \mathbf{n}_{e7} = 0 \tag{8}$$

where $\mathbf{F}_t$ is a external force acting on the traction link due to contact with the ground, m_T is the mass of the link, $\mathbf{g}$ is the gravitational acceleration vector,

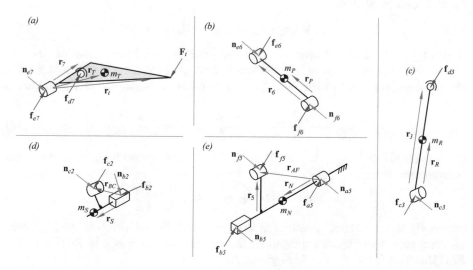

Fig. 3. Free-body diagrams of one leg of a 3RSRR wheel

$\mathbf{n}_{e7}$ is the internal moment at joint E, $\mathbf{f}_{d7}$ and $\mathbf{f}_{e7}$ are internal forces at joints D and E, respectively. The moment Eq. (8) is given with respect to point D. On the other hand, for the proximal link we have:

$$\mathbf{f}_{e6} + \mathbf{f}_{f6} + m_P \mathbf{g} = \mathbf{0} \tag{9}$$

$$\mathbf{r}_6 \times \mathbf{f}_{e6} + \mathbf{r}_P \times m_P \mathbf{g} + \mathbf{n}_{f6} = \mathbf{0} \tag{10}$$

where m_P is the mass of the proximal link, $\mathbf{n}_{f6}$ is the internal moment at joint F, $\mathbf{f}_{e6}$ and $\mathbf{f}_{f6}$ are internal forces at joints E and F, respectively. In this case, the moment Eq. (10) is given with respect to point F. Equations from (7) to (10) represent a system of 12 equations with 13 unknowns (i.e., the internal forces and moments at joints D, E and F.) The force components at joints that are orthogonal to the wheel axis can be determined by considering the static force equations in the wheel's plane. Regarding that $\mathbf{f}_{e7} = -\mathbf{f}_{e6}$, the force components at joint E in the plane $x_1 - y_1$ can be computed by the following expression:

$$^1\bar{\mathbf{f}}_{e7} = \mathbf{K}^T \begin{bmatrix} ^1\bar{\mathbf{r}}_7^T \\ ^1\bar{\mathbf{r}}_6^T \end{bmatrix}^{-1} \begin{bmatrix} (^1\bar{\mathbf{r}}_t^T - ^1\bar{\mathbf{r}}_7^T)\mathbf{K}^1\bar{\mathbf{F}}_t + (^1\bar{\mathbf{r}}_T^T - ^1\bar{\mathbf{r}}_7^T)\mathbf{K}m_T{}^1\bar{\mathbf{g}} \\ ^1\bar{\mathbf{r}}_P^T\mathbf{K}m_P{}^1\bar{\mathbf{g}} \end{bmatrix} \tag{11}$$

the symbol $^-$ over vectors means that only the first two elements are considered, and the left superscript means the reference frame in which they are measured (e.g., $^1\bar{\mathbf{f}}_{e7} = \begin{bmatrix} ^1f_{e7x} & ^1f_{e7y} \end{bmatrix}^T$ is a subvector of the tridimensional vector $^1\mathbf{f}_{e7}$.) Matrix $\mathbf{K}$ is given by:

$$\mathbf{K} = \begin{bmatrix} 0 & 1 \\ -1 & 0 \end{bmatrix} \tag{12}$$

The corresponding forces at joints E and F, $^1\bar{\mathbf{f}}_{d7} = -^1\bar{\mathbf{f}}_{e7} - ^1\bar{\mathbf{F}}_t - m_T{}^1\bar{\mathbf{g}}$ and $^1\bar{\mathbf{f}}_{f6} =^1\bar{\mathbf{f}}_{e7} - m_P{}^1\bar{\mathbf{g}}$, are obtained from Eqs. (7) and (9), respectively.

Regarding the diagram of Fig. 3c the following moment equation with respect point C can be written:

$$\mathbf{r}_3 \times \mathbf{f}_{d3} + \mathbf{r}_R \times m_R \mathbf{g} + \mathbf{n}_{c3} = \mathbf{0} \tag{13}$$

where $\mathbf{f}_{d3}$ is the internal force at spherical joint D, m_R is the mass of the rod, $\mathbf{n}_{c3}$ is the internal moment at rotational joint C. From the previous equation the force component at joint D in the z_2 axis can be determined in the following way:

$$^2f_{d3z} = \frac{^2r_{3z}{}^2f_{d3y} + ^2r_{Rz}m_R{}^2g_y - ^2r_{Ry}m_R{}^2g_z}{^2r_{3y}} \tag{14}$$

the magnitude of $^2f_{d3y}$ is in function only of the orthogonal force $^1\bar{\mathbf{f}}_{d7}$, since $^2\mathbf{f}_{d3}$ is computed by:

$$^2\mathbf{f}_{d3} = -{}^2_1\mathbf{R}^1\mathbf{f}_{d7} = -{}^2_1\mathbf{R} \begin{bmatrix} ^1\bar{\mathbf{f}}_{d7} \\ ^1f_{d7z} \end{bmatrix} \tag{15}$$

where $^2_1\mathbf{R}$ is a rotation matrix of reference frame Σ_1 with respect to Σ_2. Once all components of $^2\mathbf{f}_{d3}$ are computed, the internal moment $\mathbf{n}_{c3}$ is obtained from Eq. (13). The force at joint C is given by:

$$^2\mathbf{f}_{c3} = -{}^2\mathbf{f}_{d3} - m_R{}^2\mathbf{g} \tag{16}$$

Additionally, since $^1\mathbf{f}_{d7} = -\frac{1}{2}\mathbf{R}^2\mathbf{f}_{d3}$ the force components of $^1\mathbf{f}_{e7}$ and $^1\mathbf{f}_{f6}$ in the z_1 axis can be determined. The moments at joints E and F are computed by solving Eqs. (8) and (10) for $\mathbf{n}_{e7}$ and $\mathbf{n}_{f6}$, respectively. From Fig. 3d the force and moment at the prismatic joint B are computed with:

$$^1\mathbf{f}_{b2} = \tfrac{1}{2}\mathbf{R}(^2\mathbf{f}_{d3} + m_R{}^2\mathbf{g}) - m_S{}^1\mathbf{g} \tag{17}$$

$$^1\mathbf{n}_{b2} = \tfrac{1}{2}\mathbf{R}^2\mathbf{n}_{c3} - {}^1\mathbf{r}_S \times m_S{}^1\mathbf{g} + {}^1\mathbf{r}_{BC} \times \tfrac{1}{2}\mathbf{R}(^2\mathbf{f}_{d3} + m_R{}^2\mathbf{g}) \tag{18}$$

Finally, the force an torque required by the actuators are given by the following expressions:

$$\tau_p = {}^1f_{b2z} \tag{19}$$

$$\tau_q = - \left({}^1n_{b5z} + {}^1\bar{\mathbf{r}}_{AF}^T \mathbf{K}{}^1\bar{\mathbf{f}}_{f5} + {}^1\bar{\mathbf{r}}_N^T \mathbf{K}m_N{}^1\bar{\mathbf{g}}\right) \tag{20}$$

where m_N is the mass of the rim, the force vector at joint F is $\bar{\mathbf{f}}_{f5} = -\bar{\mathbf{f}}_{f6}$, and $^1n_{b5z}$ is the component in the z_1 axis of $^1\mathbf{n}_{b5} = -{}^1\mathbf{n}_{b2}$.

Table 1. Parameters of the links for the simulation

Parameter	Value
l_p	0.0600 m
l_2	0.0150 m
l_3	0.0700 m
l_5	0.0400 m
l_6	0.0400 m
l_7	0.0195 m
l_9	0.0900 m
γ	25.0°
m_3	0.1716 kg
m_5	0.0980 kg
m_6	0.0980 kg
m_9	0.2205 kg

A series of simulations were performed to validate the presented static force analysis. Table 1 shows the values of the parameters of the links for the simulation. The wheel and the reference frame Σ_1 are oriented in such a way the gravitational vector is given by $^1\mathbf{g} = \begin{bmatrix} 0 & -9.81\,\text{m/s}^2 & 0 \end{bmatrix}^T$. The external force is applied at the tip of the traction link and is given by $^1\mathbf{F}_t = \begin{bmatrix} 0 & -1N & 0 \end{bmatrix}^T$.

Table 2 presents the results of six different simulations using the presented formulation in Matlab. These results coincide with those obtained through multibody dynamics simulation in [11]. Each simulation considers two different values of q_r (0° and 45°) and three different values of q_p (0 cm, 1.5 cm and 3 cm). The results show the required force at the actuators (τ_r and τ_p) and the magnitude of the linear forces acting on joints C to F. As can be seen, the values of τ_p decrease as the values of q_p increase. On the other hand, the values of τ_r are larger when $q_r = 45°$, because this configuration has a larger lever arm.

Table 2. Simulation results for different values of q_r and q_p.

q_r [°]	0	0	0	45	45	45
q_p [cm]	0.0	1.5	3.0	0.0	1.5	3.0
τ_r [N-cm]	−0.6698	7.3530	12.4807	20.6077	35.5615	42.9888
τ_p [N]	−18.9525	−5.0563	−2.4253	-8.2021	−0.6467	0.2357
$|\mathbf{f}_c|$ [N]	26.2630	8.4982	6.4560	9.9923	6.2864	7.2175
$|\mathbf{f}_d|$ [N]	25.9742	7.4397	4.9989	10.1947	4.6511	5.6878
$|\mathbf{f}_e|$ [N]	25.7225	6.2440	2.9937	11.2668	1.8333	3.2547
$|\mathbf{f}_f|$ [N]	25.7228	6.1629	2.8387	11.7432	1.3321	2.8191

4   Conclusions

In this paper the static force analysis of a RSRR HeIse wheel was presented. Given the external force acting on the traction wheel the presented model determines the internal forces at the mechanism joints, the required torque at the rotational actuator and the required force in the linear actuator to achieve the static equilibrium. The results of simulations in six different positions were presented. These results coincide with those obtained through multibody dynamics simulation. The obtained expressions in this paper could be used for analysis and design purposes if the wheel moves at relatively low velocities, since the inertial forces would be negligible compared with the external forces applied to the wheel. Future work is the experimental validation of the model and its use for dimensional synthesis of the mechanism.

References

1. Burt, I.T., Papanikolopoulos, N.P.: Adjustable diameter wheel assembly, and methods and vehicles using same. US Patent 6,860,346 (2005)
2. Cordes, F., Dettmann, A., Kirchner, F.: Locomotion modes for a hybrid wheeled-leg planetary rover. In: IEEE International Conference on Robotics and Biomimetics (ROBIO), pp. 2586–2592 (2011)
3. Eich, M., Grimminger, F., Kirchner, F.: A versatile stair-climbing robot for search and rescue applications. In: 2008 IEEE International Workshop on Safety, Security and Rescue Robotics, pp. 35–40 (2008). https://doi.org/10.1109/SSRR.2008.4745874

4. Lu, D., Dong, E., Liu, C., Xu, M., Yang, J.: Design and development of a leg-wheel hybrid robot hytro-i. In: IEEE/RSJ International Conference on Intelligent Robots and Systems (IROS), pp. 6031–6036 (2013)

5. Moreno, H., Carrera, I., Garcia, J.P., Baca, J.: Ruedas heise: Familia de mecanismos para implementar ruedas hibridas de geometría variable. Revista Iberoamericana de Autom. e Informática Ind. **15**(4), 427–438 (2018)

6. Moreno, H.A., Carrera, I.G., Pamanes, J.A., Camporredondo, E.: 2 dof mechanism for a variable geometry hybrid wheel. In: Advances in Automation and Robotics Research in Latin America, pp. 6031–6036. Springer (2017). ISBN 978-3-319-54376-5

7. Nagatani, K., Kuze, M., Yoshida, K.: Development of transformable mobile robot with mechanism of variable wheel diameter. J. Robot. Mechatron. **19**, 252–253 (2007)

8. Qiao, G., Song, G., Zhang, Y., Zhang, J., Li, Z.: A wheel-legged robot with active waist joint: design, analysis, and experimental result. J. Intell. Robot. Syst. **83**, 485–502 (2016)

9. Siegwart, R., Lamon, P., Estier, T., Lauria, M., Piguet, R.: Innovative design for wheeled locomotion in rough terrain. Robot. Auton. Syst. **40**, 151–162 (2002). ISSN 0921-8890

10. Yun, S.S., Lee, J.Y., Jung, G.P., Cho, K.J.: Development of a transformable wheel actuated by soft pneumatic actuators. Int. J. Control, Autom. Syst. **15**, 36–44 (2017)

11. Zendejo, O.: Analisis de la cinematica directa y de fuerzas estaticas de una rueda heise rsrr. BS Thesis, Universidad Autonoma de Coahuila (2017)

Kinematic Analysis of a Lower Limb Rehabilitation Robot

Isela Carrera[✉], Héctor Moreno, Isidro Hernández,
and Emilio Camporredondo

Universidad Autónoma de Coahuila, Barranquilla S/N, Monclova, Coahuila, Mexico
iselacarrera@uadec.edu.mx

Abstract. There is a large number of people in Mexico and around the world with lower limb disabilities, some of them have no chance of recovery but others with adequate rehabilitation can regain their mobility. Common rehabilitation needs are the sit to stand (STS) movement and walking. This paper presents the kinematic analysis of a robot for STS and gait rehabilitation.

1 Introduction

According to the World Health Organization (WHO) over a billion people live with some form of disability [7]. The Instituto Nacional de Estadística y Geografía (INEGI) reports in 2016 that there are 7.1 million disabled people in Mexico [2]. Some physical disabilities, such as those caused by stroke or cerebral palsy, can be treated through rehabilitation exercises, which consist of constant repetitive movements to make a patient learn or relearn those movements. During a rehabilitation session, the physiotherapist employs his physical strength to perform the exercises on the patient. In recent years, different researchers have introduced designs of robotic devices to perform or assist in physical therapy. The objective of those developments is to support therapists when performing high-intensity, repetitive and/or physically demanding exercises on patients. Some advantages of rehabilitation robotic devices are the possibility of continuous assessment of the patient progress, using a variety of sensors and computational methods, and the adaptation of therapy according to each patient situation.

A lower limb disability reduces the quality of life of an affected person since it hinders or impedes mobility and various activities of daily living. Common rehabilitation needs for this kind of disability are the sit to stand movement and walking. In some rehabilitation exercises, the physiotherapist supports the patient's weight, which is a physically demanding task that could be performed using robotic devices. There are few devices that have been proposed in this regard. The rehabilitation of the sit to stand movement involves several phases where an adequate trajectory must be performed by the joints of ankle, knee, hip and trunk. In [5], it is presented an assistant walker that helps to perform the correct sit to stand movement and continues with assistance during the gait. This robot makes the correct movements of the trunk however, all force is carried

A. Martínez et al. (Eds.): LACAR 2019, LNNS 112, pp. 72–78, 2020.
https://doi.org/10.1007/978-3-030-40309-6_8

on the trunk and on the arms. Another rehabilitator can reproduce the proper movement of all involved joints and make the patient reach vertical position, it contains three intensity states and uses only one actuator, this device only rehabilitates the sit to stand movement, but does not continue with the gait [6].

On the other hand, there are some robotic devices that have been designed for gait rehabilitation. Walkaround is a robot that supports the patient's body and is intended to assist during functional electrical therapy of walking in individuals with hemiplegia [8]. In [4] the WHERE I and II rehabilitation robots are described. Those robots can be programmed to follow a training path and implement methods for tracking the movements of the user through sensors. SAM and SAM-Y (for children) are robotic walkers that support the patient during gait therapy [1]. Lokomat is a robotic system that supports patients while their legs are attached to robotic legs that assist with basic walking functions. This device includes a treadmill on which the walking is performed [3].

This paper presents the kinematic analysis of a rehabilitation robot for lower limb disabilities. The proposed device is intended to assist in therapy in order to improve functional mobility of patients in activities like sit to stand transfers, sitting, standing and walking. The following section describes the design of the mechanism, and after that, the kinematic analyses of position and velocity are obtained.

2 Position Analysis

Figure 1 shows a model of the Papalotl robot. The robot consist of a mobile platform and a backrest with body-weight support (BWS). The mobile platform

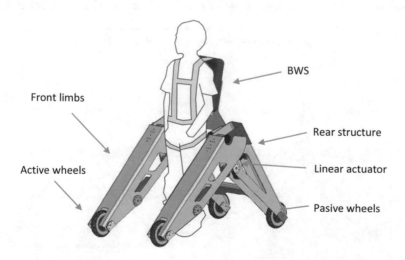

Fig. 1. Structure of the Papalotl rehabilitation robot

is composed of a rear structure and a pair of front limbs. The front limbs can be extended/flexed with respect to the rear structure by means of a pair linear actuators. The BWS is elevated when the front limbs of the mobile platform are flexed. The patient is supported by the BWS using a harness. A linear actuator controls the orientation of the BWS with respect to the rear structure. The horizontal translation of mobile platform is controlled by means of the frontal actuated wheels. The coordination of the actuators of the limbs, BWS, and front wheels permits to perform the sit to stand movement.

In this problem the task variables are the hip position $(x_h - y_h)$ and the trunk angle (ρ). The joint variables are the length of the linear actuators of the limbs (q_l) and BWS (q_r), and the position of the frontal wheel (q_x) with respect of a reference point. The solution of the inverse kinematics (i.e., determining q_l, q_l and q_l given x_h, y_h and ρ) is featured below.

The geometric parameters and kinematic variables of the mechanism are presented in Fig. 2. Consider that the value of y_h is known, the angle of the rear structure can be obtained as follows:

$$\theta_1 = \arcsin\left(\frac{y_h - R_w}{l_1}\right) \tag{1}$$

where l_1 is the length of the rear structure and R_w is the radius of the wheels.

From Fig. 2 the following sum of vectors is obtained:

$$\mathbf{r}_1 - \mathbf{r}_2 - \mathbf{r}_3 = \mathbf{0} \tag{2}$$

the unknown variables in the previous equation are x_f, that represents the distance between the wheels, and θ_2 , that represents the orientation of the front extremities. A quadratic equation in x_f can be obtained, and its solution is:

$$x_f = l_1 \cos\theta_1 \pm \sqrt{l_1^2 \cos^2\theta_1 + l_2^2 - l_1^2} \tag{3}$$

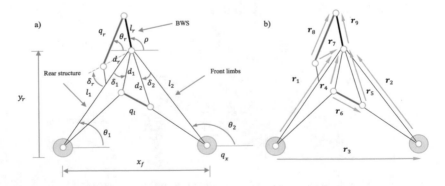

Fig. 2. Kinematic scheme of the robot: (a) Geometric parameters (b) Considered vectors for the analysis

On the other hand, the angle θ_2 is obtained as follows:

$$\theta_2 = \arccos\left(\frac{\pm\sqrt{l_1\cos^2\theta_1 + l_2^2 - l_1^2}}{l_2}\right) \tag{4}$$

In order to determine the value of the joint variable of the frontal limbs linear actuators q_l, consider the following sum of vectors in Fig. 2b:

$$\mathbf{r}_4 - \mathbf{r}_5 - \mathbf{r}_6 = 0 \tag{5}$$

the position and orientation of vector $\mathbf{r}_6$ are given by:

$$q_l = \pm\sqrt{d_1^2 + d_2^2 - 2d_1d_2\cos(\theta_1 - \theta_2 + \delta_1 + \delta_2)} \tag{6}$$

and

$$\theta_l = \arctan\left(\frac{d_1\sin(\theta_1 + \delta_1) - d_2\sin(\theta_2 - \delta_2)}{d_1\cos(\theta_1 + \delta_1) - d_2\cos(\theta_2 - \delta_2)}\right) \tag{7}$$

where δ_1, δ_2, d_1 and d_2 define the position of rotational joints (that connect the linear actuator) in the rear structure and front limbs.

On the other hand, the value of the joint variable of the BWS linear actuator can be obtained considering the following vector loop equation:

$$\mathbf{r}_7 - \mathbf{r}_8 + \mathbf{r}_9 = 0 \tag{8}$$

given the orientation of the rear structure θ_1 and the required angle of the trunk ρ, the extension and orientation of the linear actuator are obtained as follows:

$$q_r = \pm\sqrt{l_r^2 + d_r^2 + 2l_rd_r\cos(\rho - \theta_1 + \delta_r)} \tag{9}$$

and

$$\theta_r = \arctan\left(\frac{l_r\sin\rho + d_r\sin(\theta_1 - \delta_r)}{l_r\cos\rho + d_r\cos(\theta_1 - \delta_r)}\right) \tag{10}$$

where d_r and δ_r defines the position of the rotational joint that connects the rear structure and BWS linear actuator, as can be seen in Fig. 2.

Finally, the joint variable q_x is given by the following expression:

$$q_x = x_h - l_2\cos\theta_2 \tag{11}$$

Below the velocity analysis is presented, and then a simulation using the obtained kinematic model.

3   Velocity Analysis

In this section the velocity inverse kinematics is presented. Given the hip linear velocity $(\dot{x}_h - \dot{y}_h)$ and BWS angular velocity $(\dot{\rho})$, the velocities of the joint variables $(\dot{q}_l, \dot{q}_r$ and $\dot{q}_x)$ are determined. Differentiating Eq. (1) with respect to time, the angular velocity of the rear structure can be obtained:

$$\omega_1 = \frac{\dot{y}_h}{l_1\cos\theta_1} \tag{12}$$

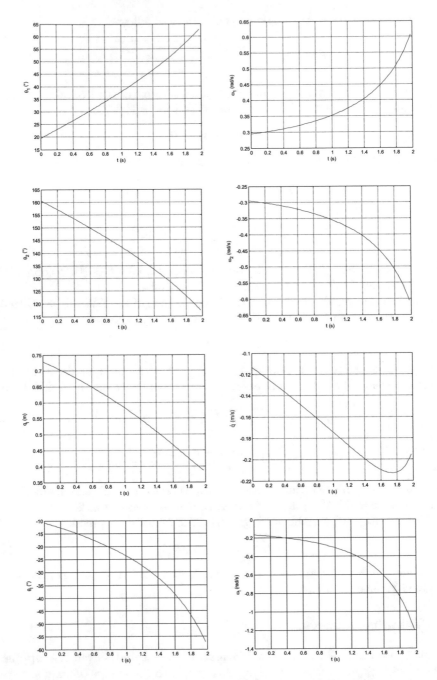

Fig. 3. (a) Orientation of the rear structure, (b) Angular velocity of the rear structure (c) Orientation of the frontal limbs (d) Angular velocity of the frontal limbs (e) Extension of the linear actuator (f) Linear velocity of the linear actuator (g) Orientation of the linear actuator (h) Angular velocity of the linear actuator

Table 1. Geometric parameters of the mechanism for simulation

Parameter	Valor
l_1	0.90 m
l_2	0.90 m
d_1	0.25 m
d_2	0.60 m
δ_1	20°
δ_2	10°
R_w	0.10 m

The angular velocity of the front limbs is obtained taking the derivative of Eq. (2) with respect to time:

$$\omega_2 = \frac{l_1 \cos \theta_1}{l_2 \cos \theta_2} \omega_1 \tag{13}$$

On the other hand, the linear velocity $\dot{x}_f$ is given by the following expression:

$$\dot{x}_f = (l_2 \cos \theta_1 \tan \theta_2 - l_1 \sin \theta_1)\,\omega_1 \tag{14}$$

The next part of the analysis is solved by taking the derivative of Eq. (5). The linear and angular velocities of the front limbs linear actuators are the following:

$$\dot{q}_l = -V_1 cos\theta_l + V_2 sin\theta_l \tag{15}$$

and

$$\omega_l = \frac{V_1 \sin \theta_1 + V_2 \cos \theta_2}{q_1} \tag{16}$$

where $V_1 = d_1 \sin(\theta_1 + \delta_1)\omega_1 - d_2 \sin(\theta_2 - \delta_2)\omega_2$ and $V_2 = d_1 \cos(\theta_1 + \delta_1)\omega_1 - d_2 \sin(\theta_2 - \delta_2)\omega_2$. Following the same procedure above, the derivative of the vectorial sum in Eq. (8) consists in a pair of equations whose solution is given by:

$$\dot{q}_r = -U_1 \cos \theta_r + U_2 \sin \theta_r \tag{17}$$

and

$$\omega_r = \frac{U_1 \sin \theta_r + U_2 \cos \theta_r}{q_r} \tag{18}$$

where $U_1 = d_r \sin(\theta_1 - \delta_r)\omega_1 + l_r \sin \rho \dot{\rho}$ and $U_2 = d_r \cos(\theta_1 - \delta_r)\omega_1 + l_r \cos \rho \dot{\rho}$. Finally, the joint variable velocity $\dot{q}_x$ is given by the following expression:

$$\dot{q}_x = \dot{x}_h + l_2 \sin \theta_2 \omega_2 \tag{19}$$

A simulation of the kinematic model of the robot is presented below. The geometric parameters of the mechanism are shown in Table 1. A constant velocity motion was considered for the simulation. The movement consists of raising the hip from an initial position $y_h = 0.4\,$m to the final position of $y_h = 0.9\,$m. The value of x_h is given in such a way the value of q_r is constant. The orientation of the BSW is constant and given by $\rho = \pi/2$. The motion is performed in $2\,$s. Figure 3 shows the values of the variables θ_1, θ_2, θ_l and q_l, and the their time derivatives. The results were validated in graphic form and through numerical differentiation.

4 Conclusions

This paper presented the kinematic analysis of a rehabilitation robot for lower limb disabilities. This device is intended to assist in therapy in order to improve functional mobility of patients in activities like sit to stand transfers, sitting, standing and walking. The solution of the inverse kinematics was presented. In this problem, the hip position and the trunk angle were considered as the task coordinates. The joint variables of the mechanism are the length of the linear actuators and the position of the frontal wheel with respect of a reference point. A simulation of the robot kinematics was presented and the results were validated in graphic form and through numerical differentiation. The kinematical model presented in this work will be used to control the rehabilitation robot that is under construction.

References

1. Goddar Space Flight Center, N.: Goddards Cable-Compliant Joint Technology Gets Patients Up and Walking with SAM (2008). http://ipp.gsfc.nasa.gov/SS-SAM.html. [Internet; accesso 22-Abril-2009]
2. INEGI: La discapacidad en mexico, datos al 2014. Technical report, Instituto Nacional de Estadistica y Geografia, Mexico (2016)
3. Jezernik, S., Colombo, G., Keller, T., Frueh, H., Morari, M.: Robotic orthosis loko-mat: a rehabilitation and research tool. Neuromodul.: Technol. Neural Interface 6(2), 108–115 (2003). https://doi.org/10.1046/j.1525-1403.2003.03017.x. https://onlinelibrary.wiley.com/doi/abs/10.1046/j.1525-1403.2003.03017.x
4. KapHo, S., JuJang, L.: The development of two mobile gait rehabilitation. IEEE Trans. Neural Syst. Rehabil. Eng. 17(2), 156–166 (2009)
5. Kawazoe, S., Chugo, D., Yokota, S., Hashimoto, H., Katayama, T., Mizuta, Y., Koujina, A.: Development of standing assistive walker for domestic use. In: 2017 IEEE International Conference on Industrial Technology (ICIT), pp. 1455–1460 (2017). https://doi.org/10.1109/ICIT.2017.7915580
6. Matjacic, Z., Zadravec, M., Oblak, J.: Sit-to-stand trainer: an apparatus fortraining "normal-like" sit to stand movement. IEEE Trans. Neural Syst. Rehabil. Eng. 24(6), 639–649 (2016). https://doi.org/10.1109/TNSRE.2015.2442621
7. OMS: Informe mundial sobre discapacidad. Technical report, Organizacion Mundial de la Salud (2011)
8. Veg, A., Popovic, D.B.: Walkaround: mobile balance support for therapy of walking. IEEE Trans. Neural Syst. Rehabil. Eng. 16(3), 264–269 (2008)

Overwatch-M System: Implementation of Bayesian Statistics for Assessment of Sensorimotor Control

Juan Martinez[1], José Baca[2]([✉]), Luis Rodolfo Garcia Carrillo[2],
and Scott A. King[1]

[1] Computer Science Department, Texas A&M University-Corpus Christi,
6300 Ocean Dr., Corpus Christi, TX, USA
jmartinez91@islander.tamucc.edu
[2] Department of Engineering, Texas A&M University-Corpus Christi,
6300 Ocean Dr., Corpus Christi, TX, USA
jose.baca@tamucc.edu

Abstract. This work introduces a novel strategy that combines a Bayesian Probabilistic Theory with Mixed Reality (MR) to assess user's sensorimotor control. The purpose of this system is to estimate where does the user beliefs its hand is located from sensory information after performing a set of short MR tests. Thence, to accomplish this goal, the system applies a Bayesian approach to adjust the likelihood of a test with prior evidence. Hence, the system collects spatial information about the user's hand position in a 3D space. After data collection, the data gets structured into visual (azimuth) and proprioceptive (depth) senses. Then, the probability distribution of these two senses gets analyzed using Bayesian Statistics to adjust the user's belief about the position of the hand given these two senses. With this approach, we intend to determine a framework to collect sensory information in an MR environment and use a statistical framework to study motor performance.

1 Introduction

Sensorimotor control is a nervous system mechanism within the human body that receives sensory stimuli from its environment, it transmits, processes, and integrates the neural signal as motor command outputs to accomplish a goal [15]. As people age, daily life activities such as walking, cooking, showering, manipulating objects, etc., become difficult tasks to achieve due to a reduction in their functional performance [1]. As a result, many older adults are at risk of accidents or injuries causing an individual to relinquish their independence. Estimations show that in the United States during 2014, 28.7% of adults of 65 years or older fell, with an approximated cost of \$31.3 billion to Medicare [5]. Thus, motor performance deficit is a severe problem caused by a dysfunction of the central and peripheral nervous system. One of the aspects of this disorder is the impaired coordination [16], such as eye-hand coordination. In neurophysiology,

A. Martínez et al. (Eds.): LACAR 2019, LNNS 112, pp. 79–91, 2020.
https://doi.org/10.1007/978-3-030-40309-6_9

the study to model eye-hand coordination is essential to determine the spatial position of the hands concerning the body [1]. This coordination is vital since it integrates two critical senses, vision and proprioception, which are used by the Central Nervous System (CNS) for motor planning [4]. Proprioception is a sense that provides spatial information of different limbs to CNS; it encodes information about all the joint angles between hand and rest of the body [4]. Therefore, a reduction in the transmission of these senses through the CNS can alter neuromuscular control [15]. Different experiments have been developed to acquire sensorimotor information; for example, observational cross-sectional and matching tasks [3,15]. The matching task experiment calculates the precision of the spatial probability distribution of the localization of proprioception and visual senses [3]. For example, proprioception localization test allows a user to point to a visual target with his unseen hand. The exercise is repeated over time to obtain a spatial probability distribution of the hand's proprioceptive and visual localization of the target [3]. However, the sensory system is susceptible to variability and noise which limits the precision of our senses, therefore, computational principles for sensorimotor control has to be explored and implemented to reduce uncertainty. Techniques like Bayesian methods reduce sensory uncertainty because our brain's perceptual representation and Bayesian theory, share similar system of conditioning probability for unknown parameters [2,7].

In this work, we present a novel strategy that combines a machine learning technique based on Bayesian theory, with Mixed Reality (MR) to assess user's sensorimotor control and impact to functional performance. Basic sensorimotor virtual tests, short in duration and easy to implement, guide the user through a series of activities while monitoring user's interaction. The Bayesian model analyzes data sets from the tests and determines the probability of precision of the hand position given sensory information from vision and proprioception. The advantages of this strategy rely on its ability to combine prior and observed user's interaction within basic tests.

2   Bayesian Theory and Mixed Reality

A Bayesian network is as a graphical model that represents conditional independencies from a set of random variables [9]. The nodes in the network represent the variables on the domain connected in pairs by arcs representing the direct dependencies between them [8]. The relational significance between variables gets calculated by conditional probability distributions associated with each node, which prohibits direct cycles. Additionally, Bayesian Networks allows joining the Bayesian statistical analysis to reduce noise from the sensory system and sensory probabilistic precision with graphical models [17]. The advantages of this method reside in its capacity to represent conditional dependencies among all variables, understand a problem domain, and it permits to consolidate prior knowledge with likelihood data [17]. However, this method fails to relate new links for new observed experiences because time is an essential factor for motor planning. Therefore, a Dynamic Bayesian Network (DBN) can counteract the flaws of the static system since it is a model that changes over time [17]. Additionally, DBN models have been previously used to develop risk assessment programs. In the medical field, DBN is a tool used to provide prognosis and diagnosis of diseases like Cervical Cancer and Coronary Heart Disease [12,13].

Mixed reality (MR), is the combination of both real and virtual worlds to produce new environments and visualizations where physical and digital objects co-exist and interact in real time. While pure augmented reality (AR) overlays virtual objects on the real-world environment, MR not just overlays but anchors virtual objects to the real world and allows a user to interact with virtual objects. The benefits of this technology rely on its ability to reduce the cognitive load and increase user's interest in the training [6]. Cognitive load describes the total mental effort used by the memory to perform a new task using useful instructions [18]. Therefore, MR is a platform that conveys information through load elements, but these loads have a low extraneous cognitive load compared to 2D methods [6]. In other words, it simplifies the information being presented to the user because it shows an immersive virtual environment in a compact and realistic mode. As a result, it reduces the amount of mental energy used to learn a new task because the user receives depth cues and physical objects interaction catalyzing its performance in training [6]. Also, the use of this technology is not prone to accidents since the training is performed in a controlled environment, which makes it safe for inexperienced users to perform any physical test [11]. Another advantage is that an assessment of performance can be recorded in real-time to analyze trainees' sensory information [11].

3   Overwatch-M System

Overwatch-M System enables an objective assessment of user's functional performance via Mixed Reality tests. Bayesian model enhances the system by measuring key user's interaction parameters and tracking changes in functional performance that could lead to sensorimotor problems.

3.1   Basic Sensorimotor Tests via Mixed Reality

A set of basic virtual tests, short in duration and easy to implement, were developed with the objective to assess user's interaction during execution of task. We have selected Microsoft HoloLens, as platform, to develop MR activities or tests, as shown in Fig. 1. Each virtual test consists of eight essential components that are used to monitor interaction, i.e., training selection, manipulating object, path, time guide, window time frame, nodes and initial and final spheres.

The collected information consists of a node number, a Boolean to detect if the *time guide* was moving in forward or backward direction, X, Y, and Z coordinates of the Hololens' spatial coordinate system, the repetition number, and timestamp.

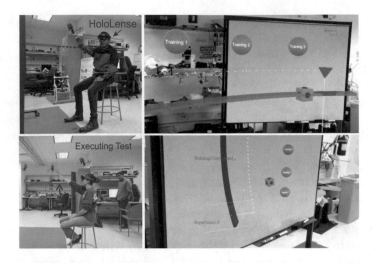

Fig. 1. Overwatch-M System: Set of basic Mixed Reality tests, short in duration and easy to implement, were developed to assess user's interaction. Each test collects data such as objects' position error and time response.

3.2   Bayesian Statistics for Sensorimotor Control

The objective of Bayesian statistics is to determine how the observed and unobserved parameters are related. Therefore, to calculate sensory input given some motor command output Bayesian statistics take advantage of generative models [19]. The dataset for the sensory input S is given by $f(S|\theta)p(\theta)$ where θ are the observed hand states. So first, we draw the probabilities of the hand state from $p(\theta)$ and label these as $\theta_1, ..., \theta_s$. All these parameters represent the possible population that generates the observed sensory input. For each θ_j, where $j = 0, ..., s$, they represent the samples that might generate the maximum utility for model $f(S|\theta)p(\theta)$. Then, for each sampled hand state a new sensory input estimate is drawn from $f(S|\theta = \theta_j)$ and declared as $S_1, ..., S_s$. Then we find the maximum S_j that agrees with the observed sensory input. The location of this new estimate represent the new parameter of the posterior distribution [14]. Thus, with the latter, we can estimate how CNS performs optimal estimates under uncertainty [19]. As an example, in an initial state with no prior information, a likelihood model $P(S_v|\theta)_0$ for vision and $P(S_p|\theta)_0$ for proprioception, is created based on the known position of the path's nodes, as shown in Fig. 2(a), for proprioception. To estimate the maximum utility which make the hand state samples $\theta_0, ..., \theta_j$ the most likely, new models are generated. The maximum estimation from the model's samples are searched. The highest estimation provide the parameters to the distribution $P(\theta|S_p)_0$, or posterior, and the highest point in the distribution becomes the Most Likelihood Estimate MLE_0. Thus, this point estimate the most probable precision for proprioception when no prior information is available. However, if prior information is present the same generative principle applies. Nevertheless, this new estimation use prior information

$P(\theta)_0$, or the posterior distribution calculated by MLE, and the likelihood of a new test $P(S_p|\theta)_1$ from newly observed samples $\theta_0, ..., \theta_j$ as shown in Fig. 2(b). Alike MLE, the maximum point of the posterior distribution $P(\theta|S_p)_1$ estimates for the most probable precision for proprioception. This estimate is known as the Maximum A Posteriori MAP_0.

$$P(\theta|S) = \frac{P(S|\theta)P(\theta)}{P(S)} \tag{1}$$

Another vital component for posteriori estimation is the denominator $P(S)$ observed in Eq. 1. The normalizing factor, or marginal likelihood, does not modify the relative probabilities of the sensory input but normalize data probabilistically between zero and one. This process repeats at every test and the current posterior becomes the new prior for future trainings [19].

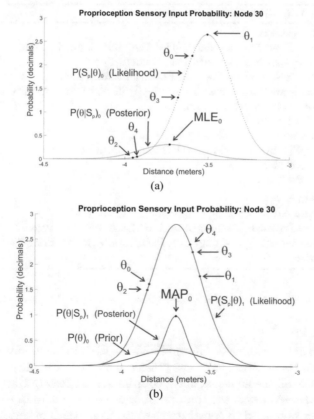

Fig. 2. (a) Maximum Likelihood Estimate for Proprioception at node 30. (b) Maximum A Posteriori for proprioception at node 30.

4   Implementation of Bayesian Theory

The Bayesian Statistics approach has been divided into four functions, *i.e.,* Main Function, Struct per Node, Posterior Model, and Maximum A Posteriori.

4.1   Main Function

The main process (Algorithm 1) requires a file of the observed hand states as input, and a prior input file if prior information exists. Once the training data gets read, it needs to organize the observed sample coordinates by their corresponding node. Following this, the posterior model parameters get returned by the Bayesian function for vision, Z, and proprioception, X. Finally, mean and standard deviation of the generated posterior model per node get stored into the Posterior file.

Algorithm 1. Main function

---

1 **Main**(*Data*, *Posterior*)
   Input: *Data*: Read file with data about the user's input training.
   Input: *Posterior*: Read file with Prior information
   Output: *Posterior*: Write file with Posterior information
   Result: Reads and structures the training collected information to estimate the
              posterior sensorimotor control
2 PriorID = Data.prior
3 node = Data.node; x = Data.X; z = Data.Z;
4 nodePath = PData.node; xPath = PData.x; zPath = PData.z;
5 **if** *PriorID > 0* **then**
6 $\quad\lfloor$ Prior = Posterior
7 TrainingData = {};
8 TrainingData=StructureData(TrainingData, node, x, z)
9 PosteriorX = Bayesian(TrainingData, PathData, PriorID, Prior, 'X')
10 PosteriorZ = Bayesian(TrainingData, PathData, PriorID, Prior, 'Z')
11 save Posterior

4.2   Struct per Node

In Algorithm 2 the sampled information's spatial information needs to be arranged by nodes. A struct datatype gets used as input and output of the function and a list of all the nodes, x, and z coordinates are feed to the function. The variable *ProgramNodes* is the total number of sample nodes in the program; a set of 40 nodes were used to construct the test. The length of the *Node* variable is the total number of sampled nodes collected from the training of approximately 200 nodes or more. Then, the input document Node gets searched by *ProgramNodes*' value. *Nodes* that match the *ProgramNodes* store the x, and z coordinates into a struct. Then, all the structs generated per *ProgramNodes* get pushed into the Data struct and returned to the main function.

Algorithm 2. Struct per Node

---

1 **StructureData**(*Data*, *Node*, *X*, *Z*)
    Input: *Data*: Struct in which the data will be organized by node.
    Input: *Node*: list of node number.
    Input: *X*: list float X coordinates
    Input: *Z*: list float Z coordinates
    Output: *Data*: Data information organized by nodes in a struct
    Result: Structures the information to be organized accordingly by nodes
2 **for** $j = 0$ to length(ProgramNodes) **do**
3   loc = 0; table = []; Xstruct = {}; Zstruct = {}; nodeVal{1} = j;
4   **for** $i = 0$ to length(Node) **do**
5   **if** Node[i] == j **then**
6   Xstruct{1,loc} = X[i];
7   Zstruct{1,loc} = Z[i];
8   loc++;
9   Data{j} = {('Node', nodeVal), ('X', Xstruct), ('Z', Zstruct)}
10 **return** Data

4.3 Posterior Model

Bayesian statistics, in Algorithm 3, consists primarily of two functions Maximum Likelihood Estimate (MLE) and Maximum A Posteriori (MAP) to find the model parameters that maximize the utility of MLE and MAP given the observed data. The difference between these two functions is the fact that the MAP requires prior information and MLE does not. *MarginalLikelihood* is a normalizing factor of the sum of all the probabilities where MLE or MAP got found. The new posterior model gets generated with the optimal parameters and gets normalized to provide a probabilistic result within zero and one. Then, the normalized posterior parameters get returned to the main function.

4.4 Maximum A Posteriori

The MAP function in Algorithm 4, as stated before, finds the parameters of the model where the MAP gets maximized from the observed information like hand state samples, likelihood model, and prior model. The search got performed by searching at every millimeter over a distance of 5 m; a delta of 0.001 by 2500 steps in a positive and negative direction from the likelihood mean. The result of each search gets stored in a list. After searching through the entire search space, the maximum MAP gets found from the list with a maximum function. The index location of the MAP from the list is then used to retrieve the Mean and Marginal Likelihood. The posterior model standard deviation gets calculated using the equation for several observations with a normal prior [10].

Algorithm 3. Posterior Model

---

1 **Bayesian**(*Data*, *PriorID*, *Prior*, *Sense*)
   Input: *Data*: Struct of the user's input training.
   Input: *PriorID*: Training test number.
   Input: *Prior*: List with the prior statistics.
   Input: *Sense*: Character of the input sense
   Output: *Posterior*: Struct with statistics of the calculated posterior
   Result: Calculate the posterior probability of the training using two generative
        models.
2 **for** $j = 0$ to length(ProgramNodes) **do**
3   User = Data$\{j\}$.Sense;
4   **if** *PriorID* $== 0$ **then**
5   mle = MLE(User);
6   PosteriorMean = mle[0]; PosteriorStd = mle[1];
7   MarginalLikelihood = mle[2];
8   Posterior = normpdf(PosteriorSamples, PosteriorMean, PosteriorStd);
9   Posterior = Posterior/MarginalLikelihood;
10   **else**
11   **if** *Sense* $=$ 'X' **then**
12   s = 1
13   **else**
14   s = 2
15   Prior[0] = Prior$\{s,0\}$[j]; Prior[1] = Prior$\{s,1\}$[j];
16   Likelihood[0] = Mean(User); Likelihood[1] = Std(User);
17   map = MAP(User, Likelihood, Prior)
18   PosteriorMean = map[0]; PosteriorStd = map[1];
19   MarginalLikelihood = map[2];
20   Posterior = normpdf(PosteriorSamples, PosteriorMean, PosteriorStd);
21   Posterior = Posterior/MarginalLikelihood;
22   P$\{s,0\}$[j] = Mean(Posterior); P$\{s,1\}$[j] = Std(Posterior)
23   **return** P;

---

5 Experimental Results

The approach has been implemented to test the sensorimotor function, eye-hand coordination. Overwatch-M system via Hololens is capable of simplifying the spatial surface of the real-world into a triangle mesh anchored into a spatial cartesian coordinate system in meters. The origin of this coordinate system is the initial head position and orientation when a program gets initiated. Also, as the user traverses around a space, Hololens collects new spatial information about the environment, as shown in Fig. 3.

For the experiment, five users performed same five tests with three repetitions per test. As a first step, we ask the user to complete a MR tutorial to familiarize with the system (user should understand how to interact with the interface). Following after, user sits down to set user's head position as the origin of the

Algorithm 4. Maximum A Posteriori (MAP)

---

1   **MPA**(*Data*, *Likelihood*, *Prior*)
    Input: *Data*: List of user's hand states.
    Input: *Likelihood*: List with Likelihood Statistics.
    Input: *Prior*: List with Prior Statistics.
    Output: *Posterior*: List with Posterior Statistics
    Result: Search for the Maximum A Posteriori using likelihood and prior information by brute-force.

2   m = 1; count = 0;

3   MarginalLikelihood = 0; PriorMean = Prior[0]; PriorStd = Prior[1];

4   LikelihoodMean = Likelihood[0]; LikelihoodStd = Likelihood[1];

5   delta = 0.001; lklN = LikelihoodMean; lklP = LikelihoodMean;

6   **for** *step = 0 to 2500* **do**

7   MeanNegative = lklN-pstd;

8   **for** *p = 0 to length(Data)* **do**

9   list(p) = normpdf(Data(p), MeanNegative, LikelihoodStd) * normpdf(lklN, PriorMean, PriorStd);

10   MarginalLikelihood = list(p) + MarginalLikelihood;

11   MAPE = list(p) * m; m = MAPE;

12   MAPEList[count][0] = MAPE; MAPEList[count][1] = MeanNegative;
         MAPEList[count][2] = MarginalLikelihood; lklN = MeanNegative; m = 1; count++;

13   **for** *step = 0 to 2500* **do**

14   MeanPositive = lklP+pstd;

15   **for** *p = 0 to length(Data)* **do**

16   list(p) = normpdf(Data(p), MeanPositive, LikelihoodStd) * normpdf(lklP, PriorMean, PriorStd);

17   MarginalLikelihood = list(p) + MarginalLikelihood;

18   MAPE = list(p) * m; m = MAPE;

19   MAPEList[count][0] = MAPE; MAPEList[count][1] = MeanPositive;
         MAPEList[count][2] = MarginalLikelihood; lklN = MeanNegative; m = 1; count++;

20   **for** *m = 0 to length(MAPEList)* **do**

21   idx = max(MAPEList[m][0])

22   MAP[0] = MAPEList[idx][1];

23   MAP[1] = $sqrt(((PriorStd^2)^{-1} + (LikelihoodStd^2/length(Data)^{-1})^{-1}))$;

24   MAP[2] = MAPEList[idx][2]

25   **return** MAP;

---

world. Otherwise, user might perform poorly since he might lose perception of the origin in which the program gets anchored. Consequently, the user must tap and drag the manipulating object to select one of the tests or trainings. For instance, training one requires three repetitions at a slow speed of approximately 28 cm per second with approximately 5 m to cover in the MR World. If training is completed flawlessly, it means it last approximately 2 min; otherwise it will run until the program completes the repetitions. Depending on the training and accuracy of the user, number of samples points in every node might vary, *e.g.*, if user moves the manipulating object away from the path, path turns red color, while the time guide is at node n the direction of the training will change to a retrospective mode, so instead of moving to n^{+1} it will move to n^{-1} towards $n^{initial}$. If this event happens frequently nodes between node n and $n^{initial}$ will contain more sample points than nodes between n and n^{final}. In this experiment, the motion of the repetitions is horizontal; right outstretch

Fig. 3. User grabs virtual object and tries to match object's centroid with cross sectional position of the *Time Guide* and the *Path*. Each test contains different speeds and number of repetitions. After completing the test, information is stored and analyzed by the Bayesian model.

arm moving towards your left shoulder, or left outstretch arm moving away from the body. This motion forces the user to stretch their arm from the initial to final sphere position, and always maintaining their hands within HoloLens's view (head has to rotate simultaneously). Otherwise, the Time Guide will pause. Additionally, a counter shows the number of completed repetitions. When all the repetitions are completed, user can stop dragging and holding the Manipulating Object and exit the program with a bloom gesture.

Experimental results show that the Bayesian technique provides a true quantitative estimate value for functional performance. Figure 4(a) displays user 1 at an initial stage (MLE_0), for proprioception input, with a low precision. This is because the method estimates the users' most likely model from the user's sampled information using optimal likelihood distribution $P(S_p|\theta)_0$. Hence, MLE adjusts the user's statistics to the users' optimal level; which in this case differs by a considering amount from the optimal model. Once optimized, the maximum posteriori gets estimated, increasing the proprioception input precision at MAP_1. After this, the maximum posteriori model adjusts accordingly to the user's sensory input with every new test i.e., from test 0 to test 5, as shown in Fig. 4(b) and (c) for proprioception and vision, respectively. The results of the maximum posteriors from five different tests seem to ascertain a positive correlation. From this observational analysis, it can be perceived that the range

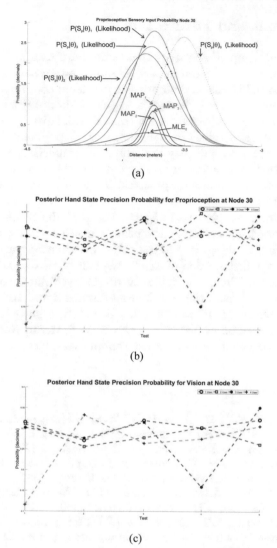

(a)

(b)

(c)

Fig. 4. (a) Proprioception sensory input probabilities distribution of five tests for user 1 (b) Maximum posterior distribution probability for vision and proprioception of five different users.

of functional performance between most of individuals, reaches similar values as number of tests increases. User 5 is a particular user that displays critical changes in functional performance. Results shows potential bad eye-hand coordination. As a matter of fact, this user was tired due to poor rest when performing the test. Hence, generating a much diverging performance in relationship with that of the other users.

6 Conclusions and Future Work

In this work, we developed a strategy that combines Bayesian Theory with MR to asses sensorimotor control and impact on motor performance. To validate our approach, we selected the sensorimotor case, eye-hand coordination. Within the MR test, the user interacted through a series of short tests and data sets of spatial information were collected for analysis. Results display the probability of precision of the hand state given the sensory input. Experimental results from the eye-hand coordination case, display that the level of precision remains relatively constant given the sensory information from the previous to the current test for node 30. Users reach a certain precision level that remains almost constant through all five tests.

For future work, we plan to integrate the probability distribution of the hand position for vision and proprioception. In other words, the distribution for both senses will be integrated into a single distribution per node. Following this integration, the distribution of each node will be mixed to generate a multimodal distribution. Then, the similarity of observed multimodal distribution will be compared with the prior multimodal distribution. This comparison will aid determine a pattern of motor control. Ultimately, a pattern of probabilistic similarity will be used as evidence in a Dynamic Bayesian Network to predict loss of coordination conditioned by other random variables.

References

1. Baca, J., Ambati, M.S., Dasgupta, P., Mukherjee, M.: A modular robotic system for assessment and exercise of human movement. In: Advances in Automation and Robotics Research in Latin America, pp. 61–70. Springer (2017)
2. Bays, P.M., Wolpert, D.M.: Computational principles of sensorimotor control that minimize uncertainty and variability. J. Physiol. **578**(2), 387–396 (2007)
3. van Beers, R.J., Sittig, A.C., van der Gon, J.J.D.: The precision of proprioceptive position sense. Exp. Brain Res. **122**(4), 367–377 (1998)
4. van Beers, R.J., Sittig, A.C., van der Gon, J.J.D.: Integration of proprioceptive and visual position-information: an experimentally supported model. J. Neurophysiol. **81**(3), 1355–1364 (1999)
5. Bergen, G., Stevens, M.R., Burns, E.R.: Falls and fall injuries among adults aged ≥65 years—united states, 2014. In: MMWR. Morbidity and Mortality Weekly Report, vol. 65, no. 37, pp. 993–998 (2016)
6. Hoe, Z.Y., Lee, I.J., Chen, C.H., Chang, K.P.: Using an augmented reality-based training system to promote spatial visualization ability for the elderly. Univ. Access Inf. Soc. **18**, 327–342 (2017)
7. Knill, D.C., Pouget, A.: The Bayesian brain: the role of uncertainty in neural coding and computation. Trends Neurosci. **27**(12), 712–719 (2004)
8. Korb, K.B., Nicholson, A.E.: Bayesian Artificial Intelligence. Chapman and Hall/CRC, Boca Raton (2003). (Chapman &Hall/CRC Computer Science & Data Analysis)

9. Labatut, V., Pastor, J., Ruff, S., Démonet, J.F., Celsis, P.: Cerebral modeling and dynamic bayesian networks. Artif. Intell. Med. **30**(2), 119–139 (2004). Artificial Intelligence in Neuroimaging: Four Challenges to Improve Interpretation of Brain Images

10. Lee, P.M.: Bayesian Statistics: An Introduction (Arnold Publication). Hodder Education Publishers (2004)

11. Lee, S., Lee, J., Lee, A., Park, N., Lee, S., Song, S., Seo, A., Lee, H., Kim, J.I., Eom, K.: Augmented reality intravenous injection simulator based 3d medical imaging for veterinary medicine. Vet. J. **196**(2), 197–202 (2013)

12. Marshall Austin, R., Onisko, A., Druzdzel, M.: The pittsburgh cervical cancer screening model a risk assessment tool. Arch. Pathol. Lab. Med. **134**, 744–50 (2010)

13. Orphanou, K., Stassopoulou, A., Keravnou, E.: DBN-extended: a dynamic bayesian network model extended with temporal abstractions for coronary heart disease prognosis. IEEE J. Biomed. Health Inform. **20**(3), 944–952 (2016)

14. Rubin, D.B.: Bayesianly justifiable and relevant frequency calculations for the applied statistician. Ann. Stat. **12**(4), 1151–1172 (1984)

15. Röijezon, U., Faleij, R., Karvelis, P., Georgoulas, G., Nikolakopoulos, G.: A new clinical test for sensorimotor function of the hand development and preliminary validation. BMC Musculoskelet. Disord. **18**(1), 407 (2017)

16. Seidler, R.D., Bernard, J.A., Burutolu, T.B., Fling, B.W., Gordon, M.T., Gwin, J.T., Kwak, Y., Lipps, D.B.: Motor control and aging: links to age-related brain structural, functional, and biochemical effects. Neurosci. Biobehav. Rev. **34**(5), 721–733 (2010)

17. Shangari, T.A., Falahi, M., Bakouie, F., Gharibzadeh, S.: Multisensory integration using dynamical bayesian networks. Front. Computat. Neurosci. **9**, 58 (2015)

18. Sweller, J., van Merrienboer, J.J.G., Paas, F.G.W.C.: Cognitive architecture and instructional design. Educ. Psychol. Rev. **10**(3), 251–296 (1998)

19. Wolpert, D.M.: Probabilistic models in human sensorimotor control. Hum. Mov. Sci. **26**(4), 511–524 (2007)

Deep Neural Network-Inspired Approach for Human Gesture-Triggered Control Actions Applied to Unmanned Aircraft Systems

Gabriel Alexis Guijarro Reyes[1]([✉]), Juan Martinez[1],
Luis Rodolfo Garcia Carrillo[2], Ignacio Rubio Scola[2], José Baca[2],
and Scott A. King[3]

[1] Unmanned Systems Laboratory (TAMUCC-USL),
Department of Computing Sciences, Texas A&M University - Corpus Christi,
6300 Ocean Drive, Unit 5824, Corpus Christi, TX 78412-5824, USA
{gguijarroreyes,jmartinez91}@islander.tamucc.edu
[2] Department of Engineering, Texas A&M University - Corpus Christi,
6300 Ocean Drive, Corpus Christi, TX, USA
{luis.garcia,ignacio.rubioscola,jose.baca}@tamucc.edu
[3] Department of Computing Sciences, Texas A&M University - Corpus Christi,
6300 Ocean Drive, Unit 5824, Corpus Christi, TX 78412-5824, USA
scott.king@tamucc.edu

Abstract. Deep Neural Networks have positively impacted the development of autonomous navigation strategies, allowing the integration of object detection capabilities to the vast array of already existing control methods. In this project, a Deep Neural Network is implemented to distinguish particular hand gestures, as well as user's arm movement, in order to trigger a specific activity of an Unmanned Aircraft System. We present real-time experimental results of how this technique is applied, making use of a Natural User Interface to control a the autonomous aircraft.

1 Introduction

The use of Natural User Interfaces (NUI) facilitate the integration of state-of-the-art technology with particular activities of a human user, in a transparent and easily understandable way. The potential of NUI extends its use to users with disabilities such as impairment of language or proper sensorimotor control. From speech recognition to computer vision, multiple complementary techniques can be used to develop an effective NUI. The collected user data is then processed making use of information coming from the analysis of historic data, or with intelligent recognition techniques such as Machine Learning (ML) and Deep Neural Networks (DNN).

Typically, a NUI relying on computer vision will deal with effective detection and classification of Objects of Interest (OoI), which are a critical element for the

A. Martínez et al. (Eds.): LACAR 2019, LNNS 112, pp. 92–111, 2020.
https://doi.org/10.1007/978-3-030-40309-6_10

Fig. 1. The experimental setup proposed to control a UAS using hand gestures. A RGBD camera connected to a ground station computer processes the gestures and position of the user's hand to create specific commands. This data is sent via a wireless link to the UAS. A Motion Capture System is used to measure the position and orientation of the UAS inside the experimental area.

decision making process. Automating and ensuring the efficacy of this process is one of the main challenges faced nowadays by NUI developers. The main goal of our research is to use computer vision and DNN to identify hand/arm gestures of a human user, and use this data to trigger real-time control actions of an Unmanned Aircraft System (UAS). The usage of DNN for object recognition may be computationally intensive to be performed directly onboard the UAS, but can be easily implemented on a ground station computer. For this reason, our approach makes use of an off-board camera connected to a personal computer. The visual information is processed and converted into control commands, which are sent to the UAS wireless. Figure 1 displays the scenario that describes the problem addressed in this article.

1.1 Related Work

A basic idea about how a DNN can be implemented to control a UAS is presented in [6]. The experimental setup used YOLOv2 [11] and a conventional USB digital camera as the unique source of data. The tracked body parts were the head and hands, from which specific detected gestures determined discrete commands such as take-off, landing, ascend, and decent. Additionally, the authors in [5] proposed a solution where DNN are used to estimate the position and speed of the UAS in a location without GPS coverage. Cloud computing is implemented in combination with Region Convolutional Neural Networks (R-CNNs) to recognize objects.

In [10], DNNs are used to detect the user and track the user pose. In this approach, different points of view of the user are provided by means of multiple micro UAS with on-board computation, and signals from additional sensors such as GPS, barometer, and inertial measurement units (IMUs) are also included in the data fusion.

1.2 Solutions Based on Tracking with Depth Cameras

The implementation of depth cameras extends the two dimensional imaging data with a third dimensional information. Along these lines, the authors in [13] presented an implementation of the Microsoft® Kinect™ camera, where the posture of the limbs and torso of the user triggered particular behaviors of a UAS. The interaction between Microsoft® Kinect™ and an UAS was also explored in [7]. Human gestures were obtained by detecting the user's skeleton, and the goal was to create a natural and consistent set of commands to control a vehicle. The authors pointed out that the results were affected due to the delay that was present in the system.

The authors in [14] proposed another controller based on the Microsoft® Kinect™. The project relies on a fusion of data from different modalities, which involved sound recognition and the implementation of simultaneous gestures. The last feature proved to be critical to accomplish acrobatic maneuvers, if needed.

1.3 Main Contributions

The research framework explored in this paper proposes a novel automated NUI which helps with the execution of certain UAS tasks. Our approach makes use of an RGBD camera in combination with a DNN, to detect specific human body positions and hand gestures. This information is later used in an event-oriented approach to trigger the UAS tasks. The proposed approach is novel in the sense that it makes use of discrete commands (hand gestures) and continuous commands (arm motion) simultaneously. These hybrid commands are ultimately used for defining the position and behaviour of the UAS during a real-time mission. This approach is set to work within indoors environments, where there is a controlled illumination along with the usage of a commercial off-the-shelf RGB camera.

The rest of this manuscript is organized as follows. Section 2 describes the problem to be solved; Sect. 3 introduces the proposed solution based on NUI for triggering UAS actions; Sect. 4 presents experimental results, followed by a brief discussion. Finally, conclusions and future directions of this research work are presented in Sect. 5.

2 Problem Formulation

A user requires to pilot an UAS remotely that is found in a controlled space, where a system is able to determine the exact location of this vehicle at anytime. Inconveniently, the subject does not know how to pilot a device of this type, since it lacks of time and resources to learn how to make use of a radio controller whose design can be overwhelming for those who just start using it.

However, in the controlled space there is a RGBD camera capable to adapt a natural interface to control this vehicle which is simple to use but with limited amount of functions and most of the time difficult to memorize.

In order to simplify the commands to control an UAS, an approach using DNN is proposed, where the system is instructed to interpret colored images from the camera. Simultaneously, the depth data tracks the user's human body joints. Being the right hand selected to move a virtual goal, which conditions the UAS to move only if a specific gesture gets made. Figure 1 shows an typical scenario of the problem addressed in this article.

2.1 Assumptions

Assumptions present in the problem are as follow:

- The UAS can follow a virtual goal manipulated via an external interface
- The UAS can starts the experiment from a random position in the proposed volume of operation
- The real world coordinate system must match with the virtual world created by the user interface
- The UAS can change the heading using selected gestures
- The UAS is controlled via a NUI
- The UAS has to react to commands given by gestures
- The depth camera is able to acquire a perfect location of the user

2.2 UAS Model

To address the problem under consideration, the quad-rotor dynamic model described in [3] is adopted. Where the Euler-Lagrange approach sets the coordinates of the quad-rotor (expressed in Eq. 1).

$$q = (x, y, z, \psi, \theta, \phi) \in \mathbb{R}^6 \qquad (1)$$

where $\xi = (x, y, z) \in \mathbb{R}^3$ corresponds to the position of the rotor-craft in the inertial frame $\mathcal{I}$, $\eta = (\psi, \theta, \phi) \in \mathbb{R}^3$ is the quad-rotor's orientation, ψ points to yaw angle in z-axis, θ belongs to pitch angle in y-axis, finally ϕ concern to the roll angle in x-axis.

Also, a rotation matrix shifted 90°, used in earlier work [4] gets applied into operations due to the experimental requirements present in the test-bed and the capture space.

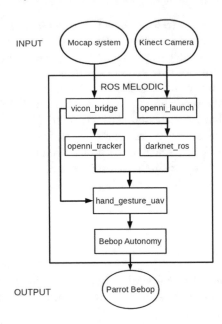

Fig. 2. Description of the ROS environment.

3   Real-Time Implementation

The platform ROS Melodic is used to integrate the different modules that involve hardware and software resources (Fig. 2).

The implementation of the experiments is divided into three sections for a better understanding (Fig. 3):

- Motion Capture System
- NUI
- UAS Control

3.1   Motion Capture System

A Motion Capture System is used to keep track of the UAS. This is done via VICON® DataStream SDK, that allows communication between the system and any other devices in a local network. The system was set to a frequency of 100 Hz. This virtual world was constrained to $1.82\,\text{m} \times 1.67\,\text{m} \times 1.70\,\text{m}$ (as shown in Fig. 4). This process is handled by node vicon_bridge, where data obtained from the SDK is transformed into suitable information to ROS, that is later used by the control node.

To match the capture space system coordinates and the Microsoft® Kinect™ created world, the following calculation is made:

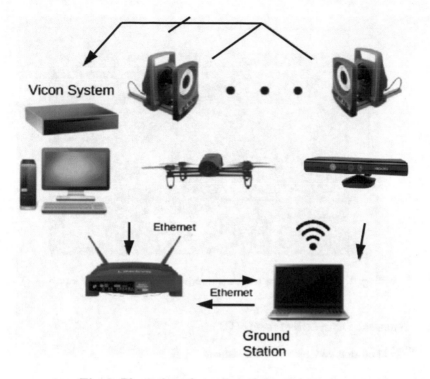

Fig. 3. Physical configuration of the environment

$$\mathbf{U}^t = \begin{bmatrix} x_{\mathrm{h}} \\ y_{\mathrm{h}} \\ z_{\mathrm{h}} \end{bmatrix} = R(\alpha\gamma) \times \xi \tag{2}$$

with

$$R(\alpha\gamma) = \begin{bmatrix} c_\alpha & -s_\alpha & 0 \\ c_\gamma s_\alpha & c_\gamma c_\alpha & -s_\gamma \\ s_\gamma s_\alpha & s_\gamma c_\alpha & c_\alpha \end{bmatrix} \tag{3}$$

Here, $\mathbf{U}^t$ corresponds to the vector that contains the desired tri-dimensional position located inside the capture space, $R(\alpha\gamma)$ is the rotation matrix that matches the world coordinate system, α is the angle of rotation in the z-axis $(90°)$, γ corresponds to the angle in X-axis $(90°)$, c stands for cos and s represents to sin.

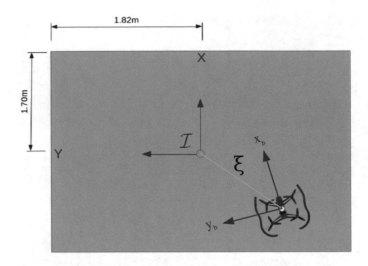

Fig. 4. Capture space where the experiments were effectuated.

3.2   Natural User Interface (NUI)

The NUI is conformed by the following:

- The NUI was created exploiting the capabilities of a Microsoft® Kinect® camera device, which used a pre-modeled detector of human bodies, those are captured and interpreted in a coordinate system. The position of each body member can be used to describe a command that responds accordingly.
- Camera device counts with different resolutions (640 * 480 px @ 30fps and 1920 * 1080 px @ 10fps) [9], these resolutions directly affects the speed of frames delivered back to the computer. Moreover, the horizontal and vertical Field of Vision (FOV) is 57° and 45° respectively. Due to the nature of this implementation, the chosen selected mode was the fast frame rate and low resolution (Fig. 5).
- Due to hardware constraints, NUI computation occurs at Ground Station (GS).
- In a parallel process, the DNN is capable of recognizing the first four letters of the American Sign Language (ASL). The database to train the DNN came from gestures captured from the RGB captured previous to the test using Fine Tuning. The total amount of pictures captured are 5600, where about 15% is used to evaluate the accuracy of the predictions made by the neural network.

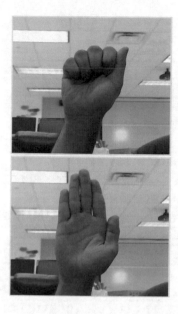 

Fig. 5. Gestures captured with a RGB Camera, from top-left to bottom-right, "A", "B", "C", "D".

- The chosen object detection system was YOLOv3 [12], where the algorithm demonstrates to perform its operations in real-time and consequently suitable to capture and process the human gestures fast enough, where remaining limitations come from the hardware itself [6]. Based on recommendations and tests performed on different hardware by [12], the variant YOLOv3-tiny (see Table 1) is applied to the experiments.
- The model produces a signal each time a gesture is detected. According with the gesture, an action is performed by the UAS, resulting in a change of the virtual goal and behaviour, once the neural network notices the gestures.
- The recollected signals are transferred to the control system using [1].

A graphical interface was created to give feedback to the user and illustrate the outcome of each movement once gesture is recognized. The interface consists of a window deployed on the screen, where a point indicates the tracking of the user's right hand. The perception of depth is represented by the size of this point, which is more prominent as the hand gets closer to the camera. A visualization of this interface can be seen in Fig. 7. The additional cameras register motion effectuated by a UAS, being this their only task to perform, no other procedure was applied to the image streams coming from those devices (Fig. 6).

In Fig. 2 is shown the process handled by nodes `openni_launch`, `openni_tracker` and `darknet_ros`. Where first node is responsible for obtaining RGB images from the camera, and second node for handling the tracking of

Table 1. Structure YOLOv3-tiny utilized to recognize the hand gestures

	Layer	Filters	Size	Input	Output	
0	conv	16	$3 \times 3/1$	$416 \times 416 \times 3$	$416 \times 416 \times 16$	0.150 BFLOPs
1	max		$2 \times 2/2$	$416 \times 416 \times 16$	$208 \times 208 \times 16$	
2	conv	32	$3 \times 3/1$	$208 \times 208 \times 16$	$208 \times 208 \times 32$	0.399 BFLOPs
3	max		$2 \times 2/2$	$208 \times 208 \times 32$	$104 \times 104 \times 32$	
4	conv	64	$3 \times 3/1$	$104 \times 104 \times 32$	$104 \times 104 \times 64$	0.399 BFLOPs
5	max		$2 \times 2/2$	$104 \times 104 \times 64$	$52 \times 52 \times 64$	
6	conv	128	$3 \times 3/1$	$52 \times 52 \times 64$	$52 \times 52 \times 128$	0.399 BFLOPs
7	max		$2 \times 2/2$	$52 \times 52 \times 128$	$26 \times 26 \times 128$	
8	conv	256	$3 \times 3/1$	$26 \times 26 \times 128$	$26 \times 26 \times 256$	0.399 BFLOPs
9	max		$2 \times 2/2$	$26 \times 26 \times 256$	$13 \times 13 \times 256$	
10	conv	512	$3 \times 3/1$	$13 \times 13 \times 256$	$13 \times 13 \times 512$	0.399 BFLOPs
11	max		$2 \times 2/2$	$13 \times 13 \times 512$	$13 \times 13 \times 512$	
12	conv	1024	$3 \times 3/1$	$13 \times 13 \times 512$	$13 \times 13 \times 1024$	1.595 BFLOPs
13	conv	256	$1 \times 1/1$	$13 \times 13 \times 1024$	$13 \times 13 \times 256$	0.089 BFLOPs
14	conv	512	$3 \times 3/1$	$13 \times 13 \times 256$	$13 \times 13 \times 512$	0.399 BFLOPs
15	conv	27	$1 \times 1/1$	$13 \times 13 \times 512$	$13 \times 13 \times 27$	0.005 BFLOPs
16	detection					
17	route	13				
18	conv	128	$1 \times 1/1$	$13 \times 13 \times 256$	$13 \times 13 \times 128$	0.011 BFLOPs
19	upsample		$\times 2$	$13 \times 13 \times 128$	$26 \times 26 \times 128$	
20	route	19, 8				
21	conv	256	$3 \times 3/1$	$26 \times 26 \times 384$	$26 \times 26 \times 256$	1.196 BFLOPs
22	conv	27	$1 \times 1/1$	$26 \times 26 \times 256$	$26 \times 26 \times 27$	0.009 BFLOPs
23	detection					

user's extremities by using the data contained in depth frames. Finally, the third node contains the DNN that used the provided RGB images. The second and third node deliver their output to the control node (hand_gesture_uav), where second node delivers the position of selected extremities to track (this position is relative to the origin of the camera). The third node determines the existence of any gesture and directs its results to the control node. Also, it handles the creation, maintenance and suspension of the user interface.

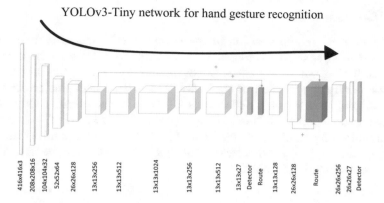

Fig. 6. Visual representation of network structure used by this methodology. It starts from a image 416×416 and finishes with 26×26 image. Each one of the white layers represents a combination of a convolutional and max layers.

Fig. 7. Graphical Interface displaying detection and recognition of two gestures represented by letter B and C, respectively.

3.3   UAS Control

Starting from a random position in the defined world, the UAS had as a goal to follow the virtual point created by the program. This goal is possible due to the NUI that converts goals into commands for the UAS. Also, two other gestures modify the heading of the UAS. The virtual goal was created by the depth camera, then the obtained data is adjusted to the world generated by the motion capture system. Where a Proportional Derivative (PD) controller was used to match the adjusted virtual goal. The PD controller handled the motion in Pitch, Roll and Yaw, where each one had different constant values. These values are displayed in Table 2.

Table 2. Values used in the PD Controller

Movement	KP	KD
Pitch & Roll	0.85	0.6
Altitude	3.5	0.0001
Heading	0.02	0.005

The control of the UAS is described in Algorithm 1. It displays the process to act every time a gesture is detected by the DNN. Where `gesture` is the received user's input, which correspond to A, B, C, or D. The ψ is the heading angle of the UAS and the term `tg` contains current target to be followed by the vehicle.

Algorithm 1. Control commands triggered by gestures
Result: Action to perform by the UAS
if $gesture = $ "A" **then**
| $\psi = \psi - 0.1$
else
  if $gesture = $ "B" **then**
  | $tg = NULL$
  else
    if $gesture = $ "C" **then**
    | $tg = Operations$
    else
      if $gesture = $ "D" **then**
      | $\psi = \psi + 0.1$
      end
    end
  end
end

The gesture assigned to letter A corresponds to an incremental rotation movement in a counterclockwise direction, B sets a $NULL$ value that commands the

UAS to hover. C by the other hand, performs the operations shown in Eqs. (4), (5) and (6), being the only command capable of updating any movement of the UAS. Finally, D corresponds to a clockwise direction.

$$x = (x_h - x_t) - x_o \tag{4}$$
$$y = y_h - y_t \tag{5}$$
$$z = z_h - z_t \tag{6}$$

Where x, y and z are the calculated position for the UAS. x_h, x_t and x_o are the coordinates in the x-axis of the right hand, torso and offset on this axis. The offset is made for user's convenience, where the zero in x-axis is displaced in front and the minimum value is reached as the right hand is at the same level as the torso. Otherwise, the user must perform anti-natural movements displacing the right hand further than the torso in direction to user's back, this potentially may cause some injuries. Meanwhile, y_h and y_t are the coordinates in the y-axis of the right hand and torso. Finally z_h and z_t are the coordinates in the z-axis.

The node in charge of this process is hand_gesture_uav, where the process creates commands to be performed by the UAS. In this step, calculations are made to obtain the location of the target using the coordinates from the camera and the mo-cap system. At the end of each iteration, the node sends the result to bebop autonomy, where the command builder creates the required commands to control the UAS.

4   Experimental Results

In this section, the technical description and results are presented using the proposed approach. A detailed description of this test-bed can be seen at [8]. To achieve an homogenized environment, the system was built in the same network. A total of 19 experiments (considering all four gestures) were executed and analyzed.

During the experiments the UAS automatically takes-off once it starts, it hovers in an initial position until a user is detected. The system starts at the moment a user appears in front of the camera and makes an initialization gesture, which is a psi pose, that resembles to the Greek letter ψ [2]. The body of the user is tracked at all times, although more than one user can fit into the system, only one is allowed to control the UAS. This is achieved by using the modified version of openni_tracker, where node accepts only one user at a time. The experiment continues until it is stopped by a Stop signal given by software or hardware via a joystick. The execution of the experiments occurred inside ROS Melodic where the environment was shared between different devices closing the gap between different communication protocols, with the usage of TCP/IP networking at a frequency of at least 100 Hz.

Ground Station specifications:

- Intel® Core i7-8750H CPU @ 2.20 GHz (12)

- 8 GB RAM DDR4
- WiFi and Ethernet connectivity
- GeForce® GTX 1050 Ti/PCIe/SSE2 (4 GB VRAM)
- Ubuntu 18.04.1 LTS
- ROS Melodic

The YOLOv3 object detection system was compiled alongside with CUDA, and the variant used is the Tiny version. The compilation in CUDA that enables operations on a GPU is critical, since most of its benefits come from this configuration [12]. The operational parameters used by Tiny-YOLOv3 are displayed in Table 3.

Table 3. Parameters utilized by the detector

Model	Threshold	Classes
Tiny - YOLOv3	0.50	'A', 'B', 'C', 'D'

Data to be captured:

- Position and orientation of the UAS
- Position of tracked hand
- Times at a gesture was captured
- Probability of detected gestures

For practical purposes, the report of results was split into two parts, where each type of command, discrete and continuous, is explored in detail. The discrete commands enclose movements to modify the heading of a UAS, allowing and denying its movement. Continuous commands correspond to the position of the tracked hand.

4.1   Testing Discrete Commands

To test the discrete commands, the gestures were registered according to their presence compared with the time that it occurred along with the probability of detection per presented gesture and the class of said gesture. At Fig. 8(a), it can be seen the tracking applied to the user's right hand, the green point is placed over the hand, regardless gesture being detected or not. This is related to user's feedback about current state and position of a virtual goal. Another example can be visualized in Fig. 8(b), where a C is detected, allowing to update the virtual goal and moving the UAS consequently.

To evaluate the performance of this approach, tests of 35 s each, were divided into five intervals (seven seconds per interval), where the first interval (0–7 s) is for preparation and is ignored. Each of the gestures were made using the sequence A (interval: 7–14), B (interval: 14–21), C (interval: 21–28) and D (interval: 28–35). The results in Fig. 9 shows that the letter A only appears at the second

Fig. 8. (a) Hand tracking enabled once the `psi` pose was made by the user, the green circle was updating the virtual goal represented by the position of the circle and the numbers above the interface. (b) Hand tracking along with gesture detection following the sign and position of user's hand. It can be appreciated the user giving a letter `C` as input to the system, this updates the virtual goal's position, moving the UAS as indicated.

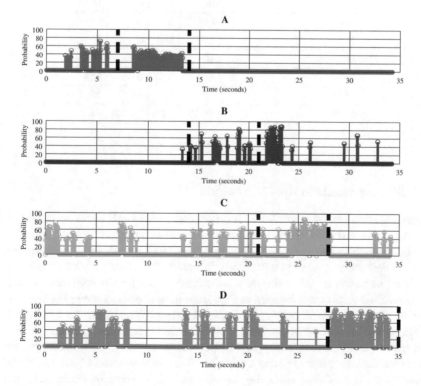

Fig. 9. Predictions of each gesture captured through the experiment 19.

interval as expected. Similarly happens for the rest of the gestures, however, one can notice that those gestures (B, C and D) also appear in all intervals. The reason for this is due to the resolution of the captured images. We consider this as noise, and should be treated in the future. This noise increases the difficulty to recognize some of the gestures. Nevertheless, based on the number of detections within corresponding intervals and despite the fact that the system is using only raw data from the DNN, i.e., there is not post-processing data or any type of filter, one can say that the approach gives acceptable results.

To evaluate success rate of proposed approach, we performed different tests and analyzed data obtained from 180 samples where each of the signals (A, B, C and D) were captured replicating similar movement. The gesture created by the hand, was moved side to side within camera's view, changing depth simultaneously. Success rate for each of the signals are summarized in Table 4.

Table 4. Evaluation of success rate (S.R.) obtained from the test of 180 samples per signal, which includes every false detection of other signals as well as the omission of any signal.

Executed signals (Input)	Ground truth (Output)			
	A	B	C	D
A	141	0	0	3
B	0	144	0	0
C	19	16	142	17
D	8	0	0	114
Miss	12	11	38	46
S. R.	78.3%	80%	78.9%	63.4%

4.2   Testing Continuous Commands

To test the continuous commands, the position of an UAS and the hand tracked were captured and compared. Both are continuously tracked by the depth camera and the motion capture system. However, the UAS should not move unless the sign C is detected by the system. Once detected, it updates a virtual goal, which the UAS has to follow using the calculations are made by the controller to achieve said goal. This action can be seen in Fig. 10, where the UAS moved according last updated virtual goal. Video experiments at https://youtu.be/3ryXO9mXObI.

The experiments were performed according to a simple routine where the hand moved from side to side, then the proximity in the x-axis, finally, a circular movement that involves x and y axes. The altitude was fixed to 1.10 m for better appreciation in contrast with the background. The error of these tests can be seen in Table 5.

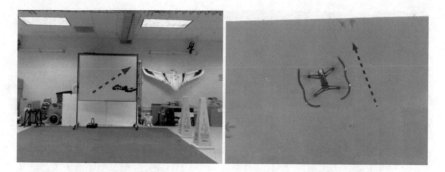

Fig. 10. The picture shows that the UAS acts according to user's commands.

Table 5. Results obtained from each test.

Test	Error in X (meters)	Error in Y (meters)	Error in Z (meters)	Error in heading (deg)
1	−0.04442	−0.19778	0.26648	12.75955
2	0.01627	−0.03118	0.12986	10.98229
3	0.04210	−0.13143	0.22384	8.50095
4	−4.90637e−05	0.00876	0.12324	11.17096
5	0.02912	0.07831	0.15331	46.18901
6	−0.07313	0.05076	0.15753	16.98305
7	0.04036	0.01803	0.12372	114.56165
8	0.03968	−0.01426	0.15119	13.28068
9	−0.02617	0.03748	0.13700	12.96737
10	0.06947	0.05275	0.26879	19.04543
11	−0.17400	−0.10919	0.45394	23.47278
12	−0.00712	−0.01793	0.13773	10.02862
13	0.01895	0.04698	0.14770	14.07118
14	−0.03032	0.05699	0.19083	24.40826
15	−0.02045	−0.00803	0.17559	17.93577
16	−0.02271	0.15370	0.35050	47.67255
17	−0.04231	−0.02325	0.12513	12.37656
18	0.09042	−0.01981	0.19051	13.24543
19	−0.01698	0.04257	0.13860	14.77058

Finally, one of those test is shown in Figs. 11, 12, 13 and 14. According to the results, the difference of position per axis did not differ by a considerable amount compared with the virtual goal. The interactions of gestures can be visualized at the figures to determine how this action change the vehicle's behaviour, where the intervals are marked in the same fashion as the previous section, an example of this can be seen at the response related with the heading where the highest change occurs at the second and fifth interval corresponding to the gestures that change this parameter.

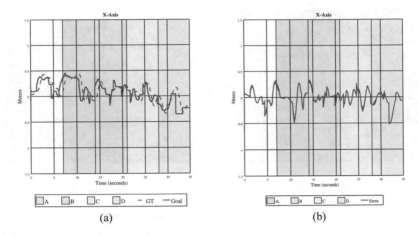

Fig. 11. (a) The average motion of ground truth and virtual goal in X-Axis as it was modified through the experiment 19. (b) Average error in X-Axis captured through the experiment 19.

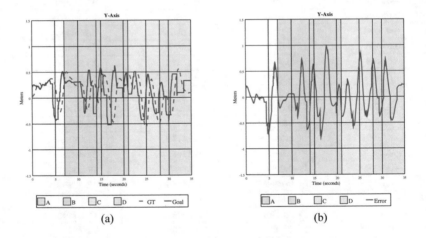

Fig. 12. (a) The average motion of ground truth and virtual goal in Y-Axis as it was modified through the experiment 19. (b) The average motion of ground truth and virtual goal in Y-Axis as it was modified through the experiment 19.

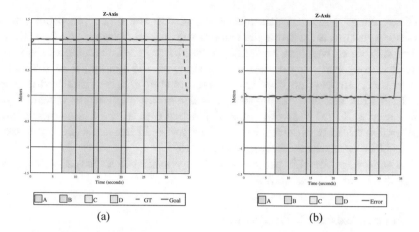

Fig. 13. (a) The average motion of ground truth and virtual goal in Z-Axis as it was modified through the experiment 19. (b) Average error in Z-Axis captured through the experiment 19.

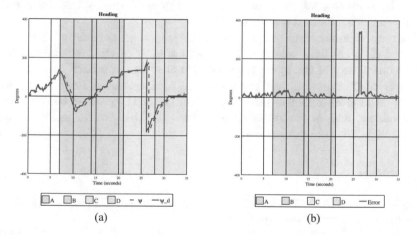

Fig. 14. (a) The average heading captured through the experiment 19. (b) Average error in heading captured through the experiment 19.

5   Conclusions

The usage of discrete and continuous commands in a NUI can help to increase the number of commands that can be used, meaning that the combination of different gestures and positions can map a larger set of possible commands, that not only can be used to direct the position and status, but also to establish simultaneous communications between a single or multiple vehicles and the user. The results showed that the resolution of the RGB camera can affect the performance of the system, since the pixels at the gesture can be distorted depending on how far is from this camera. This is also caused by the limitations of Tiny-YOLOv3 where the effectiveness is directly affected by the size of the object, in other words, if the desired gesture is far from the camera nor is well defined, the detection will not be possible or can lead to multiple readings. Mentioned faults disturb the control lecture, producing undesired reactions on the system, it is also possible to observe certain overlap between the results that can be visualized in Figs. 9, 11 and 12. That aside, it can be seen how the aircraft follow the control signals, where the axis X and Y are the most disturbed due to the noise coming from a neural network. This is an aspect that has to be addressed with filters such as Kalman filter or machine learning techniques.

The code used for this project is publicly available at https://github.com/TOTON95/Hand_Gesture_UAV

5.1 Future Research Directions

The direction of this project aims to design an intuitive, robust, and pragmatic solution that could make human-machine interaction easier and safer when working within same space or environment. The goal of this research project is to eventually remove equipment located at the ground station and have built-in control strategies located within the UAS.

Current approach could be improved by collecting more data from different users and different light conditions. Developers must provide an understandable and adequate way to approach the users. Finally, this approach could include a new layer of information over the reality (e.g., Augmented Reality and Mixed Reality), where features like previsualization of each maneuver before it is sent to the vehicle, and/or continues updates from the UAS, could facilitate control of multiple UAS and improve immersion.

References

1. Bjelonic, M.: YOLO ROS: real-time object detection for ROS (2016–2018). https://github.com/leggedrobotics/darknet_ros
2. Field, T., Liebhardt, M.: OpenNI tracker ROS driver (2012). https://github.com/ros-drivers/openni_tracker
3. Garcia Carrillo, L.R., Dzul López, A.E., Lozano, R., Peégard, C.: Quad rotorcraft control: vision-based hovering and navigation. Springer (2013)
4. Guijarro Reyes, G.A., Garcia Carrillo, L.R., Rangel, P.: A geometric control strategy for real-time coordination of multiple unmanned aircraft systems. In: 2018 International Conference on Unmanned Aircraft Systems (ICUAS) (2018). https://doi.org/10.1109/icuas.2018.8453484
5. Lee, J., Wang, J., Crandall, D., Šabanović, S., Fox, G.: Real-time, cloud-based object detection for unmanned aerial vehicles. In: 2017 First IEEE International Conference on Robotic Computing (IRC), pp. 36–43 (2017). https://doi.org/10.1109/IRC.2017.77
6. Maher, A., Li, C., Hu, H., Zhang, B.: Realtime human-UAV interaction using deep learning. In: Zhou, J., Wang, Y., Sun, Z., Xu, Y., Shen, L., Feng, J., Shan, S., Qiao, Y., Guo, Z., Yu, S. (eds.) Biometric Recognition, pp. 511–519. Springer, Cham (2017)

7. Mashood, A., Noura, H., Jawhar, I., Mohamed, N.: A gesture based kinect for quadrotor control. In: 2015 International Conference on Information and Communication Technology Research (ICTRC), pp. 298–301 (2015). https://doi.org/10.1109/ICTRC.2015.7156481
8. Muñoz Palacios, F., Espinoza Quesada, E.S., Sanahuja, G., Salazar, S., Garcia Salazar, O., Garcia Carrillo, L.R.: Test bed for applications of heterogeneous unmanned vehicles. Int. J. Adv. Robot. Syst. **14**(1) (2017). https://doi.org/10.1177/1729881416687111
9. OpenKinect: Protocol documentation (kinect) (2013). https://openkinect.org/wiki/Protocol_Documentation#RGB_Camera
10. Price, E., Lawless, G., Ludwig, R., Martinovic, I., Bülthoff, H.H., Black, M.J., Ahmad, A.: Deep neural network-based cooperative visual tracking through multiple micro aerial vehicles. IEEE Robot. Autom. Lett. **3**(4), 3193–3200 (2018). https://doi.org/10.1109/LRA.2018.2850224
11. Redmon, J., Farhadi, A.: Yolo9000: better, faster, stronger. In: 2017 IEEE Conference on Computer Vision and Pattern Recognition (CVPR) (2017). https://doi.org/10.1109/cvpr.2017.690
12. Redmon, J., Farhadi, A.: Yolov3: an incremental improvement. arXiv (2018)
13. Sanna, A., Lamberti, F., Paravati, G., Manuri, F.: A kinect-based natural interface for quadrotor control. Entertain. Comput. **4**(3), 179–186 (2013). https://doi.org/10.1016/j.entcom.2013.01.001
14. Waibel, M.: Controlling a quadrotor using kinect (2011). https://spectrum.ieee.org/automaton/robotics/robotics-software/quadrotor-interaction

Design and Simulation Analysis
of a Modular Aerial System

José Baca[1(✉)], Nohelia Jimenez[1], Kyle Winfield[1], Simone Tay[1],
Brianna Tijerina[1], Hans Baierlipp[1], Jonathan Cortez[1], and Hector Moreno[2]

[1] Department of Engineering, Texas A&M University-Corpus Christi,
6300 Ocean Dr., Corpus Christi, TX, USA
jose.baca@tamucc.edu
[2] Facultad de Ingeniería Mecánica, Universidad Autónoma de Coahuila,
Barranquilla S/N, Guadalupe, 25750 Monclova, Coah., Mexico
h_moreno@uadec.edu.mx

Abstract. This paper describes a conceptual design and simulation
analysis of a modular aerial system (MAS). This system is designed
with the purpose of independent and cooperative flight with or without payload. Properties of modularity allow the system to adapt to different tasks by adding/removing modules to/from a configuration. The
MAS module is based on a coaxial motor and a two degree-of-freedom
mechanism that transfers its weight or center of mass from one side to
another to make the module navigate around. The connector mechanism
is magnetic-based allowing the module to be attached to different metallic objects or magnets. Navigation principle, dynamic and finite element
analysis are simulated via Autodesk Inventor and MATLAB and Simscape Multibody and presented in this work.

1   Introduction

Commercial off-the-shelf unmanned aerial vehicles (UAVs) are limited by both
payload capacity and battery life while flying. Typical UAVs are either designed
to record aerial footage, or for recreational use. The capability of a standard
UAV is limited by its task-oriented design [4]. Particularly, multi-rotor systems
such as quadcopters, hexacopters and octocopters on the market possess pairs of
propellers and maneuver by varying the thrust to each propeller [5,7]. However, a
malfunction on any of their components, such as a propeller, battery, etc., could
cause significant delay or failure in any type of task or mission. Each of the
mentioned systems have different benefits and disadvantages. A quadcopter is
less expensive, small in size (compared to the rest), and great for carrying small
objects (not strong enough to carry heavy payloads). The hexacopter brings
better stability and can partially continue working even after losing a motor. It
can also fly higher than a standard quadcopter and can carry heavier payloads.
The octocopter is the strongest among them, it can fly at greater heights and
carry heavier payloads. However, it is expensive, it requires constant battery

A. Martínez et al. (Eds.): LACAR 2019, LNNS 112, pp. 112–118, 2020.
https://doi.org/10.1007/978-3-030-40309-6_11

recharge, and its size makes the system difficult to transport. Depending on the application and task, the multi-rotor system has to be designed according to specific needs either as a single task or for a potential collaborative task [6,8,9].

The modular aerial system (MAS) presented in this work attempts to bring a solution to these challenges. MAS is composed of modules of a compact size that can work in a collaborative manner with the possibility to carry different payloads according to the needs. If one of the modules present a malfunction, this module could be quickly replaced by another module and continue the task. A modular system could bring versatility, robustness and scalability to different scenarios, since modules can be re-arranged in different ways to form different configurations [1–4].

Fig. 1. Modular Aerial System concept for cargo and sensor deployment

2 Modular Aerial System (MAS)

The main element of a Modular System is the module itself. The conceptual idea of this system follows the idea of having homogeneous modules capable of executing typical tasks done by fixed wing and multi-rotor systems, such as, taking-off, hovering, navigating and carrying payloads. The scope of the project is to design and build a working prototype based on the modularity concept that could perform similar tasks as conventional systems do, while integrating the advantages of a modular system. Due to the modularity nature, several robot configurations can be assembled for fast response to different tasks such as, in urban search and rescue tasks, (e.g., safely and quickly scanning for threats inside buildings, delivering equipment, etc.), geospatial applications (e.g., deployment of sensor networks either for establishing/improving communication within restricted area or for data gathering from ground, sea, and air), etc. A Modular Aerial System (MAS) will be capable of transforming itself on-the-fly and adapt to different situations and needs, as shown in Fig. 1.

3   MAS Module

The main element of MAS, the module, is based on a coaxial contra-rotating motor (as the main source of propulsive power), a two degrees-of-freedom (DOF) mechanism or lower arm (as the mechanism that drives the robot around), and a connector mechanism that allows the module to connect to another module(s) or objects. The module is capable of navigating around by altering the position of its center of mass. This is accomplished by changing the position of the two DOF mechanism, as shown in Fig. 2.

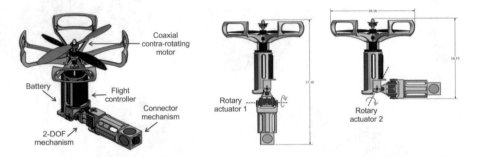

Fig. 2. MAS is based on a contra-rotating motor and a 2-DOF mechanism to navigate.

3.1   Navigation Principle

Similar to transverse weight movements in ships, the position of the center of mass of a MAS module can be modified by adjusting the 2-DOF mechanism. This adjustment causes the axis of the coaxial motor to tilt. As air flows around the module, pressure and shear stress distribution cause aerodynamic forces to act on the body in motion. These forces, lift (the strongest force acting on the module) and drag, act orthogonal when considering only the propellers. It can be observed in Fig. 3, the forces acting on the module when moving horizontally or while changing direction. Lift and thrust act in line with the flow of air through the propellers. The drag acts opposite the direction of movement, similar to friction. The resultant force, is decomposed into a horizontal (x-axis) and vertical (z-axis) component due to the inclination angle of the propellers' plane. Using Newton's second Law, it is possible to determine an estimation of the acceleration on the x-axis due to the decomposition of the resultant force divided by the mass of the prototype. Due to actuation of the acceleration on x, a resulting velocity on the x-axis allows directional control.

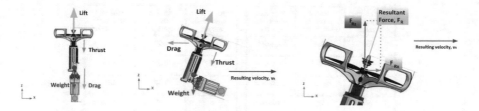

Fig. 3. Forces acting on MAS module when moving around the environment.

4   Simulation Analysis

The MAS module was designed and simulated to validate its performance. Two simulation analysis were performed, i.e., a dynamic analysis and a finite element analysis. The simulation results were used to determine final design parameters of the module's parts and how they affect module's performance.

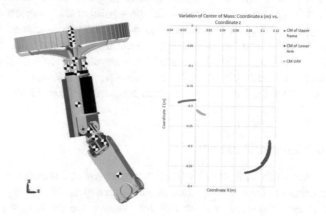

Fig. 4. Displacement of the center of mass when rotating DOF in Y-axis.

4.1   Dynamic Analysis

The MAS module, particularly, the 2-DOF mechanism or lower arm has been simulated using the program Simscape Multibody along with Simulink to verify that the lower arm can drive the module around. Simscape program was used to evaluate the change in the position of the center of mass for the upper frame and lower arm of the prototype. The input for this analysis was the actuation of the first DOF (rotary actuator 1) between a range of $\pm 45°$ from an equilibrium position where the upper frame was aligned with the z-axis. Figure 4 displays upper and lower parts of the module, alongside with the position of their corresponding center of mass (CM). Upper part's CM is displayed in blue, lower

Input rotation angle servo motors	Output rotation angle Y-axis	MAS unit movement direction	Input rotation angle servo motors	Output rotation angle Y-axis	MAS unit movement direction
-45°	12.10°	Backwards	-45°	4.96°	Left
-30°	8.28°	Backwards	-30°	3.30°	Left
-15°	4.08°	Backwards	-15°	1.52°	Left
0°	0°	Equilibrium	0°	0°	Equilibrium
15°	-5.95°	Forward	15°	-1.95°	Right
30°	-10.90°	Forward	30°	-2.90°	Right
45°	-15.50°	Forward	45°	-5.50°	Right

(a)   (b)

Fig. 5. (a) MAS displacement when rotating first DOF of the lower arm. (b) MAS displacement when rotating second DOF of the lower arm.

part's CM in red, and CM of the whole system in green color. The displacement of these points indicates a disturbance in the equilibrium position of the whole system causing a rotation in the roll angle on the propellers plane that is fundamental to the forward and backward displacements of the module. Similarly, the second DOF (rotary actuator 2) is tested to analyse system's displacement. By rotating it ±45°, the module was able to move in the left and right directions. A summary of MAS module direction with respect to both DOFs rotation can be seen in Fig. 5.

4.2   Finite Element Analysis

FEA simulations were performed with the objective of identifying the location of points for critical stress concentration and subsequently modify the design if needed. The parts that were analyzed under stress were the upper section of the module which supported the lift force of 40N that was applied as an upward vertical load. Also, the lower part of the upper section had an additional load that was applied downward on the extreme of the 2-DOF mechanism simulating the payload being lifted. The results of the thrust test by the motor, suggested that the input forces applied on the simulation could be at least 40 N.

The results from the Von Mises stress analysis, Safety factor analysis and Equivalent strain analysis (Fig. 6) indicates that the module prototype is able to withstand the stress caused by the lifting force since the coloring of the safety factor analysis shows a minimum value of 2.27 located on the contact between the upper pair of servos and the U-joint located a the center of the module. It is important to highlight that in the event of extreme forces acting on the prototype, the design is likely to fail at this joint. A summary of these simulation results are shown in Table 1.

(a)   (b)   (c)

Fig. 6. (a) Von Mises stress analysis. (b) Safety factor analysis. (c) Equivalent strain analysis.

Table 1. FEA simulation analysis - results summary

Name	Minimum	Maximum
Volume	572124 mm³	572124 mm³
Mass of Prototype (Theoretical)	1.961 kg	1.961 kg
Von Mises Stress	0 MPa	9.276 MPa
1st Principal Stress	−5.967 MPa	9.608 MPa
3rd Principal Stress	−12.868 MPa	4.675 MPa
Displacement	0 mm	0.897 mm
Safety Factor	2.270 ul	15 ul
Equivalent Strain	0 ul	3.650 E−3 ul
1st Principal Strain	−1.521 E−7 ul	2.415 E−3 ul
3rd Principal Strain	−3.937 E−3 ul	1.109 E−7 ul

5   Conclusions

In this work, a Modular Aerial System (MAS) concept was introduced, along with a simulation analysis of its dynamics and physical structure. The dimensions of the module's frame permits to fit all electrical components, as well as, components necessary to add the modularity aspects into the system. The 2-DOF mechanism or lower arm permits the module to maneuver around the X-Y plane by changing its center of mass location, and when in combination with the coaxial motor, it allows the system to maneuver around the three axes. The connector mechanism (located as end-effector) brings multiple possibilities when trying to address different tasks by creating different configurations. FEA simulations of the system proved the module could withstand 40 N of force, a value exceeding the expected forces in normal flight conditions. The dynamic simulations demonstrated the module's means of displacement by changing the location of its center of mass. By doing this, the module was able to produce movements in the forward, backward, left and right direction. As future work, the MAS module will be fabricated and tested to analyse real performance of the system.

References

1. Baca, J., Ambati, M.S., Dasgupta, P., Mukherjee, M.: A modular robotic system for assessment and exercise of human movement. In: Advances in Automation and Robotics Research in Latin America, pp. 61–70. Springer (2017)
2. Baca, J., Ferre, M., Aracil, R.: A heterogeneous modular robotic design for fast response to a diversity of tasks. Robot. Auton. Syst. **60**(4), 522–531 (2012)
3. Baca, J., Hossain, S., Dasgupta, P., Nelson, C.A., Dutta, A.: Modred: hardware design and reconfiguration planning for a high dexterity modular self-reconfigurable robot for extra-terrestrial exploration. Robot. Auton. Syst. **62**(7), 1002–1015 (2014)
4. Cardenaz, E., Ramirez-Torres, J.G.: Autonomous navigation of unmanned aerial vehicles guided by visual features of the terrain. In: 2015 12th International Conference on Electrical Engineering, Computing Science and Automatic Control (CCE). IEEE (2015)
5. Gupte, S., Mohandas, P.I.T., Conrad, J.M.: A survey of quadrotor unmanned aerial vehicles. In: 2012 Proceedings of IEEE Southeastcon, pp. 1–6 (2012)
6. Jung, S., Kim, H.: Analysis of Amazon Prime air UAV delivery service. J. Knowl. Inf. Technol. Syst. **12**(2), 253–266 (2017)
7. Liew, C.F., DeLatte, D., Takeishi, N., Yairi, T.: Recent developments in aerial robotics: a survey and prototypes overview. CoRR (2017)
8. Mohamed, N., Al-Jaroodi, J., Jawhar, I., Lazarova-Molnar, S.: A service-oriented middleware for building collaborative UAVs. J. Intell. Robot. Syst. **74**, 309–321 (2013)
9. Niedzielski, T.: Applications of unmanned aerial vehicles in geosciences: introduction. Pure Appl. Geophys. **175**(9), 3141–3144 (2018)

A Multi-Objective Genetic Algorithm Approach for Path Planning of an Underwater Vehicle Manipulator

Ilka Banfield[(⊠)] and Humberto Rodriguez

Technological University of Panama, Panama City, Panama
{ilka.banfield,humberto.rodriguez}@utp.ac.pa

Abstract. In this work the kinematic redundancy resolution based on Multi-Objective Genetic Algorithm (MOGA) has been exploited to path planning of a underwater vehicle-manipulator system (UVMS). Some objective functions are analyzed with the purpose to achieve the desired evolution of the configuration of the mobile manipulator when the position and orientation of the effector is imposed. In the proposed method, the generalized coordinates, additional kinematic constraints and relevant objectives functions are selected, in such a way that the base motion is implicitly limited while maximizing the whole system manipulability. Simulations for a ten dof vehicle-arm are developed, with the consideration of three objective functions for optimization. The results reveal that it is possible to choose among several solutions from the Pareto Front, according to the importance of each individual objective or in other way, there are an objective function that dominated the optimization process.

1 Introduction

Starting from the simple combination of a locomotion system and a manipulation system, a great variety of systems is obtained. Among the mobile manipulators, many underwater robots have also emerged to addressed the need for autonomous execution of complex task, which involves several challenging issues. In this sense, most of the research about the underwater manipulation task, such as ship hulls maintenance, pipeline weld inspection an oil or gas searching, using Underwater Vehicle Manipulator Systems (UVMS), have been focused on control strategies. Moreover, these systems are characterized by several strong drawbacks, namely, the complexity of dynamical models, external disturbances (e.g. sea currents) and limitations on the measuring sensors performance (e.g. range, accuracy and signal to noise ratio).

The best performance can be achieved, for certain tasks, by exploiting the inherently redundancy of the underwater vehicle manipulator, when solving the inverse kinematic problem as an optimization problem with specially defined objective functions. This paper presents an exploration of the kinematic control of Vehicle Manipulator Systems based on its redundancy, as a key to overcome some of the difficulties to control these systems. In this work, the objective

A. Martínez et al. (Eds.): LACAR 2019, LNNS 112, pp. 119–130, 2020.
https://doi.org/10.1007/978-3-030-40309-6_12

actions to achieve were the smoothness of joint motions, motion reduction of the base and manipulability.

With regard to the UVMS kinematic redundancy problem, the approach followed by some authors was to generate trajectories corresponding to given task, while the extra DOFs are used to assign motion without affecting the end-effector's performance. In this sense, Tang et al. [11] propose a UVMS task-priority redundancy resolution with restoring moments, optimized on accelera-tion level using a scalar potential function. In addition, they reduced the effect of restoring moments by applying gradient projection. These authors also addressed the solution to redundancy by minimization of the norm of joint's velocities using the weight pseudoinverse matrix [10].

Other work on the use of task priority approach for the coordinated motion control of UVMS was proposed by Simetti et al. [8]. Within this strategy, end-effector position and orientation is given a secondary priority objective while the high priority task is assuring each joint is within its range of motion, to keep the manipulability measure above the given positive threshold and the horizon-tal vehicle attitude in the desired rang. Cieslak et al. [4] propose an approach that combines the pseudoinverse and Damped Least-Squares (DLS) methods for calculating the Jacobian Matrix inverse by checking if the system is close to a singularity, but also supplying additional degrees of control, using a diagonal positive-definite weighting matrix. Sivčev et al. [9] presented an extensive bibli-ography about UVMS, covering research results in the field of control algorithms, including low level motion control, high level kinematic control and motion plan-ning schemes.

Generally, when considering the motion planning problems of UVMS, it is desirable that the trajectory to the goal be computed online, during motion. However, the inherent difficulty in solving this problem [7], prevents it from being solved sufficiently fast to be done online given reasonable computation resources. As a result, there are two branches of research in the area of motion planning: off-line and on-line planning. Research on off-line planning focuses on repeatable task, globally optimal solutions, completeness and overall computa-tional complexity. The challenges in on-lines planning are determining how far a solution is from the optimal and if it is bounded by an upper limit.

When the planning path is solved for a complex redundant robotic systems, one optimization method that is has gained popularity for solving complex problems in robotics is the Genetic Algorithm approach. Their performance could be superior to that classical optimization techniques and has been used successfully in robot path planning. One important drawback of Muti-objective Genetic Algorithms, MOGA's, is its on-line implementation because of nature-inspired algorithms are time-consuming and high computational cost in find-ing an optimal solution. However, compared to differential methods, (e.g. task priority) that only guarantees to be locally optimal [14], MOGA's exhibit a fast global convergency. Rodriguez et al. [6] made an exploration of the meth-ods of solving the problem of redundancy of robotic systems. They selected the MOGA Non-Sorting Genetic Algorithm (NSGA II) [5], due to the characteristics

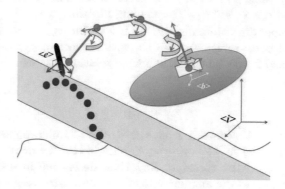

Fig. 1. Underwater vehicle manipulator schematic: coordinate systems

of the method to address multi-objective optimization problems. NSGA-II was implemented in a 5 DoF planar mobile robot, with two optimization objectives, manipulability and smoothness on joint motions. The results highlighted the need to expand the implementation (by increasing the degrees of freedom and number of objective functions) for the analysis of the selected optimization objectives. In this paper, we propose the application of NSGA-II to the trajectory generation on joint's space by searching the inverse kinematic solutions to pre-defined end effector robot path, while considering the problems of manipulability, base motion and energy consumption of a redundant UVMS. Since it is assumed that the application's environment around the robot is structured and well known to the robot, off-line motion planning is the best option. The system's model using a omnidirectional mobile-vehicle of 10 Dof is simulated in Matlab environment.

2 Kinematic Model

Underwater manipulators generally have serial-chain mechanical structure similar to industrial robot arms. There are many studies in the literature about robot kinematics that can be applied to underwater manipulators [2]. When the mobile manipulator has an holonomic platform, the manipulator modelling can be directly applied. Thus, the model the model considered in this work comprises a omnidirectional floating body with a 4 Dof manipulator was illustrated in Fig. 1. The systems is describe as a rigid body based on the mathematical model derived by Antonelli [2], using the references: $<i>$ as inertial reference frame, $<b>$ as floating body-fixed reference frame and $<e>$ as end-effector reference frame. The end effector position and orientation vector with respect $<i>$ is $\mathbf{x}_e \in \mathbb{R}^6$. Let define $\mathbf{q} = \left[\mathbf{q}_b{}^T, \mathbf{q}_{mi}{}^T\right]^T \in \mathbb{R}^n$ be the state variables of the UVMS, where $\mathbf{q}_b = \left[\eta_{b1}{}^T, \eta_{b2}{}^T\right]^T \in \mathbb{R}^6$ is the position and orientation vector of floating body vehicle in $<i>$. The vector of angular position of the manipulator's joints is $\mathbf{q}_m \in \mathbb{R}^{(n-6)}$. The vehicle position is $\eta_{b1} = [x_b, y_b, z_b]^T$ and $\dot{\eta}_{b1}$ is the corresponding time derivative. Let define $\nu_{b1} = [u_b, v_b, w_b]^T$, as the linear velocity of body-vehicle frame $<b>$ with respect to $<i>$, expressed in the body floating-fixed frame. Then the following relation between the defined linear velocities holds,

$$\nu_{b1} = {}^bR_i\dot{\eta}_{b1}, \tag{1}$$

where ${}^bR_i$ is the rotation matrix from $<i>$ to $<b>$. Euler-angles coordinates of the floating body vehicle $<b>$ in ground $<i>$ is the vector $\eta b_2 = [\phi_b, \theta_b, \psi_b]^T$ and the vector $\dot{\eta}_{b2}$ is the corresponding time derivative in $<i>$. Also, $\nu_{b2} = [p_b, q_b, r_b]^T$, is define as the angular velocity in $<b>$ with respect to $<i>$. Then ν_{b2} and $\dot{\eta}_{b2}$ are related by

$$\nu_{b2} = J^{rb}(\eta_{b2})\dot{\eta}_{b2}, \tag{2}$$

where $J^{rb}(\eta_{b2})$ is the proper Jacobian matrix of transformations for angular velocities. Now, by defining the matrix $J^b({}^b\mathbf{R}_i)$ as

$$J^b({}^b\mathbf{R}_i) = \begin{bmatrix} {}^bR_i & 0_{3x3} \\ 0_{3x3} & J_{rb} \end{bmatrix} \in \mathbb{R}_{6x6} \tag{3}$$

it is

$$\nu_b = J^b({}^b\mathbf{R}_i)\dot{\eta}_b \tag{4}$$

is the Jacobian matrix transforming the velocities (linear and angular) from $<i>$ to $<b>$. Let define $\dot{\xi} = \begin{bmatrix} {}^b\dot{\eta}_{e1}^T, {}^b\dot{\eta}_{e2}^T \end{bmatrix}$ as the velocity of vehicle-manipulator end-effector (linear and angular) in $<b>$, then,

$$\dot{\xi} = {}^b J_e(q)\dot{\mathbf{q}}_m \tag{5}$$

where ${}^b J_e(q)$ is a geometric Jacobian matrix, which maps the generalized system velocities of the end-effector's velocity in $<b>$ from $<e>$. Finally, $\dot{\zeta} = [\nu_b^T, \dot{\mathbf{q}}_m^T] \in \mathbb{R}^n$ is the velocity vector including the vehicle velocities as well as the manipulator joint velocities in $<b>$,

$$\dot{\mathbf{x}}_e = \mathbf{J}(\mathbf{q})\dot{\zeta} \tag{6}$$

where $\dot{\mathbf{x}}_e = \begin{bmatrix} {}^i\dot{\eta}_{e1}^T, {}^i\dot{\eta}_{e2}^T \end{bmatrix}$

$$\mathbf{J}(q) = \begin{bmatrix} {}^i\mathbf{R}_b & -{}^i\mathbf{R}_b(\eta_{b2})\left[\eta_{e1}\right]_\times \cdot \mathbf{J}_{rb}^{-1} & {}^i\mathbf{R}_b \cdot {}^b\mathbf{J}_{pe}(\mathbf{q}) \\ 0_{3\times3} & {}^i\mathbf{R}_b(\eta_{b2})\mathbf{J}_{rb}^{-1} & {}^i\mathbf{R}_b \cdot {}^b\mathbf{J}_{re}(\mathbf{q}) \end{bmatrix} \tag{7}$$

3   Performance Criteria

Path planning of a mobile-manipulator, refers to the task of discovering an optimal path in the joint space (or operational space) by optimizing certain objective functions. The planned path provides the task objectives that make up the current action to be executed and generates the poses for the desired position and orientation of the end-effector. Generally, the task objectives must to accomplish with physical constrains objectives, system safety, the application oriented objectives and optimization objectives [3]. The next section are considered objectives in such a way that the tasks are translated into a corresponding set of system position references (of the manipulator's joint and the linear and angular axes of the floating base).

3.1 Manipulability Analysis

As a manipulability measure, we consider the Yoshikawa index [13], which is very useful to characterize the instantaneous kinematic condition of a given system. The kinematic model of UVMS presented in Eq. 7, extends the concept of manipulability applied to mobile manipulators, defined as,

$$w = \sqrt{(det\,(JJ^T))} \tag{8}$$

In the same idea, we can extend the notion of singularity for the floating manipulator. A configuration of the UVMS is said to be singular if $rank\,(\mathbf{J}\,(q)) \neq max_q rank\,(\mathbf{J}\,(q))$. Moreover, the manipulability ellipsoids do not correspond to the inequality $\|\dot{\mathbf{q}}\| \leq 1$ by to but to $\|\dot{\zeta}\| \leq 1$, which is an important difference when also considering that base is nonholonomic. If the base is nonholonomic $w_{min} = 0$ it provides information about singular configurations, but in case of omnidirectional vehicle, the manipulability index gives a good first insight on the isotropy of the system.

The structure of the UVMS used in this work is composed of an omnidirectional floating base and a four-link serial manipulator with revolute joints, as shown in Fig. 1. The generalized coordinates of the UVMS are $\mathbf{q}_b = [x_b, y_b, z_b, \phi_b, \theta_b, \psi_b]^T$ and $\mathbf{q}_m = [q_1, q_2, q_3, q_4]^T$, base and arm respectively.

Depending on the application we may have to consider the whole system manipulability or only the robotic arm manipulability. For some task is possible that the user wants to keep the platform motionless to manipulate with the arm alone, then it would be convenient to reach the operating site in a good configuration for the arm, from a manipulation point of view. Therefore, for consideration only the robotic arm of UVMS, the kinematic model of a grounded vehicle-manipulator is used to analyze manipulability index in each point of the defined path. The manipulability surface shown in the Fig. 2(a) and (b), correspond to of the variation of the coordinates q_2 y q_3, when the remaining coordinates remain constant, with the knowledge that q_1 and q_4 do not influence the manipulability [1]. The first joint of the manipulator q_1 does not affect the manipulability of the whole system; intuitively the first joint contributes only to lateral movements of the end-effector, which are already accomplished by the omnidirectional mobile base and the last joint q_4, is determined by the end-effector orientation.

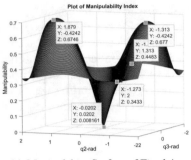

(a) Manipulabity Surface of Fixed Arm

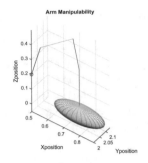

(b) Optimal Configuration Fixed Arm

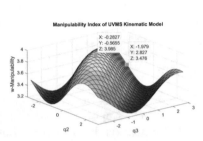

(c) Manipulabity Surface Whole System

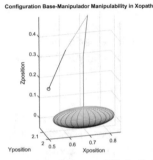

(d) Optimal Configuration Whole System

Fig. 2. Vehicle-manipulator and UVMS kinematic model for end-effector in the position initial of the path.

The effect of the mobility of the base in the index of manipulability, is measured by imposing some tasks in the operational space such as, navigation toward a particular point or performing a free floating manipulation or inspection. For this purpose and thinking in a inspection task, the imposed conditions are: the initial position of a defined path, $\mathbf{x}_e = [0.5m, 2m, 0.2m]$, only one euler angle for the end-effector is defined, θ_e, the rest are free, joints limits and the restriction in the base's coordinates $[z, roll, pitch]$ are considered. The index of manipulability is optimized using genetic algorithm. The result generated is the configuration shown in the Fig. 2. This optimal configuration corresponds $\mathbf{q}_{opt} = [0.6441m, 2.2125m, 0, 0, 0, -1.4137, -1.2373, -1.4137, -2.8274, 0.7854]$ and $w = 3.5$, under the prescribed conditions. This is the reference value of the manipulability index desired as a result of the multi-objective optimization in its competition with the other two objectives that will be defined in the next section. It can be remarked that the arm manipulability may be poor whereas the whole system keeps a better as possible measure of manipulability (compare Figs. 2a vs c and 2b vs d). While for fixed arm model the maximum manipulability reached is 0.44, in whole model the maximum manipulability in the same condition is 3.5. This underlines that the choice of the manipulability (whole system or only the arm) could be task-dependent.

3.2   Smoothness

The second function f_2, is based on error between the previous generalized position $\mathbf{q}_{old}$ and the estimated generalized position $\mathbf{q}$ in order to minimize the joint displacement along the trajectory of the UVMS, as follow

$$f_2 = \|\dot{\mathbf{q}}\| \tag{9}$$

for the implementation, two successive discrete-time samples of the joint coordinates are used,

$$f_2 = \sqrt{(\frac{\mathbf{q} - \mathbf{q}_{old}}{\Delta t})^T(\frac{\mathbf{q} - \mathbf{q}_{old}}{\Delta t})} \tag{10}$$

In a analogous manner to that the Eq. 9, the norm of the generalized accelerations is defined and it constitutes the third objective function,

$$f_3 = \|\ddot{\mathbf{q}}\|_W \tag{11}$$

but in this case, the objective function aims the minimization of weighted norm of accelerations,

$$f_3 = \ddot{\mathbf{q}}^T W \ddot{\mathbf{q}} \tag{12}$$

W is a weight matrix, that must be positive defined. In this sense, the inertia matrix of the system is selected as the weight matrix, which furthermore, by the system's weight description, allows the natural restriction of the elements of greater inertia, resulting in an energy-efficient trajectory.

These two objectives and the manipulability index are optimize to obtain continuity on the entire trajectory interval and a best energy distribution. The continuous velocity and acceleration at the via points ensure a smooth path.

4   Problem Formulation

In the present case, the task objective is that given the end-effector's sequential positions and orientations, it has to solve for the system's configurations set while satisfying kinematic constrains in a efficient and smooth manner. In order to achieve this objective, a multi-objective optimization problem was formulated for solving the inverse kinematic problem. The optimization objective functions are the maximization of Eq. 8 and the minimization of Eqs. 9 and 12, which seek the singularity avoidance (in the form of the manipulability measure), motion smoothness and efficiency by reduction of displacement, in accordance the bodies or link weights, specially for the floating body. The multicriteria optimization problem is defined as follows:

$$\begin{aligned} &min(f_1(\mathbf{q}), f_2(\dot{\mathbf{q}}), f_3(\ddot{\mathbf{q}})) \\ &s.t. \quad f(\mathbf{q}) - \mathbf{x}_d = \mathbf{0} \\ &q_i^- \leq q_i \leq q_i^+ \quad for \quad i = 1, 2, \cdots, n \end{aligned} \tag{13}$$

where $f_1(\mathbf{q}) = \frac{1}{w}$, $f(\mathbf{q})$ is the constrain vector, $\mathbf{x}_d$ is the desired pose of end-effector, q_i^+ and q_i^- denote respectively the upper and lower generalized coordinates of UVMS.

4.1   Path Design

To evaluate the objectives, the optimization problem needs a specific path. We define a representative trajectory of the end-effector in the inertial coordinate system, which is

$$x_e = 0.075t + 0.5\,[m]$$
$$y_e = -0.25x_e^3 + 2.0\,[m] \tag{14}$$
$$z_e = 0.2 + 0.1t^{\frac{1}{3}}\,[m]$$

Only one euler angle is specified for the end-effector, i.e. $\theta = \frac{\pi}{4}$, the rest is free. The path is divided into 20 segments.

4.2   NSGA-II Parameters

Path planning is an algorithmically difficult search problem. However there are very powerful, effective and robust tools for optimization problems in a wide range of applications. Under the requirement of optimized two o more objectives functions only the Evolutionary algorithms consider all the objectives, treating each objectives separately. These techniques also perform well approximating solutions to all types of problems because they ideally do not make any assumption about the underlying fitness landscape [12]. The process ended on Pareto Front which gives more than user selection. The next step is discover a global optimal solution, which is also a research topic.

The parameters of NSGA-II-are: (a)Population size = 2000, (b) Crossover function: crossover heuristic (c) Distance Crowding distance: phenotype-ratio = 0.2 and (d) Selection function: tournament.

5   Result Analysis

Figures 3, 4 and 5 show the results of generating the trajectory with the proposed models. Figure 3 is shown as part of the exploratory analysis to find an appropriate set objective functions for the task. Notice how a smooth displacement of the base is not achieved when using only objective functions f_1 and f_2. On the other hand, Figs. 4 and 5 show the results obtained when considering the three objectives functions. It is observed that under the assumed restrictions, the results are similar and it is possible to obtain an improvement in the manipulability of the arm for the kinematic control. Moreover, a lower computational requirement, fixing the base, could even be used for the online solution of many task, since the computation time is largely reduce for each degree of freedom that is not included. Pareto Fronts allow us to analyze the relationship between

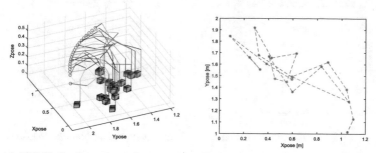

(a) UVMS Configurations. Optimization (b) Displacement of Base on X-Y Plane
of f1 and f2

Fig. 3. Results of the inverse kinematic of the systems without the consideration of
the objective f_3.

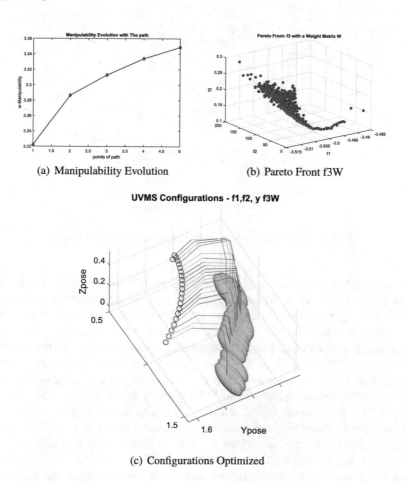

(a) Manipulability Evolution　　　　　　(b) Pareto Front f3W

(c) Configurations Optimized

Fig. 4. Floating body with arm model and f3 with a weight matrix W

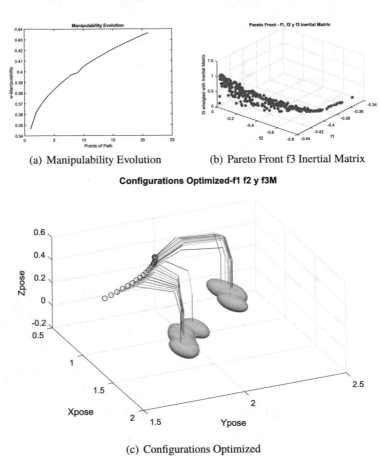

(a) Manipulability Evolution   (b) Pareto Front f3 Inertial Matrix

(c) Configurations Optimized

Fig. 5. Vehicle-Arm and f3 weighted with inertial matrix

objective functions. It is denoted that it is the manipulability function dominates in the selection of the optimum. Figures 4b and 5b, corresponds to the Pareto Fronts of the generations for the last position of the trajectory. Those Pareto Fronts presents two clusters; in the first one we have f_1 as a function of f_3 with f_2 practically constant. In the other cluster, f_1 presents a maximum limit value independent of the variations of f_2 and f_3. The above is the confirmation that the selection of these three objectives is adequate for the achievement of the objectives set. The minimization of the weighted accelerations norm, objective function f_3, allows manipulator motion with adjustable reduce motion of the base, without a predetermined task of control over the vehicle, while keeping the manipulability optimized, objective f_1. The relation that is established between the norm of the speeds and of the accelerations, with objective f_1 at its limit, allows to generate a smooth and continuous movement. Figures 4c and 5c show the configurations for each positions of the desired path and the minimal motion of the floating base without a explicit path planning for the vehicle. In the case of the use of inertial matrix of the system for the objective function f_3, Fig. 5c, the motion of the vehicle is reduced considerably and its produce a obvious change of position only when the manipulability value is not adequate.

6   Conclusion

This paper focuses on multi-objective optimization multi-objective as a tool for the exploration and solution of kinematic control problem of floating vehicle manipulators, UVMS. An analysis about the manipulability index of mobile manipulator systems and the interactions among objectives is conducted. The global search allow by NSGA-II can still offer more knowledge about the inter-action between functions and the selection of the best solution according to the task. This initial knowledge proved that is possible (at least kinematically) a indirectly controlled motion of the floating base while only the end-effector path are specified. However, although the mass of the system is included in the dynamics, its effect on the dynamical behavior of the UVMS has not been stud-ied in this work. Therefore, the next step will be to extend the analysis with the dynamic simulation of the UVMS and taking in to account drag and task forces.

References

1. Ancona, R.: Redundancy modelling and resolution for robotic mobile manipulators: a general approach. Adv. Robot. **31**(13), 706–715 (2017)
2. Antonelli, G.: Underwater Robots: Motion and Force Control of Vehicle-Manipulator Systems. Springer Tracts in Advanced Robotics. Springer, Heidelberg (2006). https://books.google.com.pa/books?id=3aAYAQAAIAAJ
3. Casalino, G., Zereik, E., Simetti, E., Torelli, S., Sperindé, A., Turetta, A.: Agility for underwater floating manipulation: task subsystem priority based control strat-egy. In: 2012 IEEE/RSJ International Conference on Intelligent Robots and Sys-tems, pp. 1772–1779 (2012). https://doi.org/10.1109/IROS.2012.6386127
4. Cieslak, P., Ridao, P., Giergiel, M.: Autonomous underwater panel operation by GIRONA500 UVMS: a practical approach to autonomous underwater manipula-tion. In: 2015 IEEE International Conference on Robotics and Automation (ICRA), pp. 529–536 (2015). https://doi.org/10.1109/ICRA.2015.7139230
5. Deb, K., Pratap, A., Agarwal, S., Meyarivan, T.: A fast and elitist multiobjective genetic algorithm: NSGA-II. IEEE Trans. Evol. Comput. **6**(2), 182–197 (2002). https://doi.org/10.1109/4235.996017
6. Rodriguez, H., Banfield, I.: Inverse kinematic multi-objetive optimization for a vehicle-arm robot system using evolutionary algorithms. In: 21st International Conference on Climbing and Walking Robots and the Support Technologies for Mobile Machines, Robotics Transforming the Future. CLAWAR. Elsevier (2018)
7. Shiller, Z.: Off-line and on-line trajectory planning **29**, 29–62 (2015). https://doi.org/10.1007/978-3-319-14705-5_2
8. Simetti, E., Casalino, G., Torelli, S., Sperinde, A., Turetta, A.: Experimental results on task priority and dynamic programming based approach to underwater floating manipulation. In: 2013 MTS/IEEE OCEANS - Bergen, pp. 1–7 (2013). https://doi.org/10.1109/OCEANS-Bergen.2013.6608016

9. Sivcev, S., Coleman, J., Omerdic, E., Dooly, G., Toal, D.: Underwater manipulators: a review. Ocean Eng. **163**, 431–450 (2018). https://doi.org/10.1016/j.oceaneng.2018.06.018

10. Tang, Q., Liang, L., Li, Y., Deng, Z., Guo, Y., Huang, H.: An energy minimized solution for solving redundancy of underwater vehicle-manipulator system based on genetic algorithm, pp. 394–401 (2017). https://doi.org/10.1007/978-3-319-61824-1_43

11. Tang, Q., Liang, L., Xie, J., Li, Y., Deng, Z.: Task-priority redundancy resolution on acceleration level for underwater vehicle-manipulator system. Int. J. Adv. Robot. Syst. **14**(4) (2017). https://doi.org/10.1177/1729881417719825

12. Vikhar, P.A.: Evolutionary algorithms: a critical review and its future prospects. In: 2016 International Conference on Global Trends in Signal Processing, Information Computing and Communication (ICGTSPICC), pp. 261–265 (2016). https://doi.org/10.1109/ICGTSPICC.2016.7955308

13. Yoshikawa, T.: Manipulability of robotic mechanisms. Int. J. Robot. Res. **4**(2), 3–9 (1985)

14. Zhang, Y., Li, J., Zhang, Z.: A time-varying coefficient-based manipulability-maximizing scheme for motion control of redundant robots subject to varying joint-velocity limits. Optim. Control Appl. Methods **34** (2013). https://doi.org/10.1002/oca.2017

A Collaborative Vacuum Tool
for Humans and Robots

Wilson Hernandez, Alvaro Hilarion, and Carol Martinez[✉]

Department of Industrial Engineering, School of Engineering,
Pontificia Universidad Javeriana, Bogotá, Colombia
carolmartinez@javeriana.edu.co

Abstract. This paper presents the design and implementation of a tool
for Human-Robot collaborative tasks. Industry 4.0 proposes a new scenario where robots can safely work in direct cooperation with humans,
within a defined workspace. These robots are called collaborative robots
or COBOTS. In this paper, we propose a tool that can be used by a robot
and/or an operator for pick and place tasks. The tool was designed to
meet certain criteria for the robot, so that it can identify it, hold it, and
perform the task. Additionally, it was designed to comply with ergonomic
parameters for the human operator. A prototype of the tool was printed
with a FusedForm 3D printer. This prototype was used for a test where
a robot and an operator will use the tool for picking plastic bottles from
a worktable. An UR3 robot was used in the test. It was equipped with
a camera in charge of identifying and estimating the position of the tool
using ArUco codes placed on the tool. Results show the functionality
of the tool for both the robot and the human operator. Additionally,
ergonomic tests provided insights to improve the handling system of the
tool for the human operator.

1 Introduction

Collaborative robots work in spaces immersed with humans. They should be safe,
easy to use, and flexible in terms of the tasks they conduct [1]. One of the most
important parts of the robots is its end effector. It allows the robot to conduct
different tasks. End effectors can be found in different styles. Suction cups, grippers (typically two or more fingers), spot-welding tools, paint sprayer, or almost
anything else that can meet the application needs. Most of the time, collaborative robotics has focused on the robot side. As Kieffer mentions in [4]: "End
users want a collaborative robot application. You can not make that if only the
robot is collaborative". Additionally in [13], it is mentioned that it is important
to develop collaborative grippers, taking into account that "they might come
into contact with the people". They must be carefully designed, so that the tool
exchange time should be short to avoid affecting production requirements. Currently, the exchange of tools is conducted in several ways [9]. The first mode is
the manual exchange. It is the simplest mode, but it has disadvantages such as

A. Martínez et al. (Eds.): LACAR 2019, LNNS 112, pp. 131–141, 2020.
https://doi.org/10.1007/978-3-030-40309-6_13

its high costs and that the operator must enter the workspace of the robot every time the tool has to be changed. The second mode is automated exchange. The exchange is done by commands using the robots control system. The automated gripper exchanges can be divided into three groups:

- multi launching grippers
- whole gripper exchange.
- clamping pads.

An example of an automatic exchanger is the one developed by Kuka [9]. It is a multifunctional end effector (MFEE) for 6 different tools, such as spindle, for sealant applications, hammer modules, and milling, among others (Fig. 1, left image). On the other hand, Fig. 1 right image shows an example of the MC-10 manual exchanger developed by ATI Industrial Automation [2]. It allows the connection of different tools in a simple way.

Fig. 1. Commercial exchanger. On the left, the automatic multifunctional end effector (MFEE) developed by Kuka [9]. On the right, the MC-10 Manual Tool Changer developed by ATI [2].

Another type of commercial grippers are the adaptive grippers. They have the advantage that they adapt to different size and shapes using simple mechanisms and simple control [9]. Some examples are shown in Fig. 2. The SRDK-UR from Universal Robots (Fig. 2, left image) is the first customizable grip system that imitates the human hands in tasks that require flexibility, dexterity, and precision [11]. The FlexShape Gripper by Festo (Fig. 2, right image) uses the chameleon's tongue principle to grip. It can grip different types of objects regardless their shape.

Industry 4.0. proposes a new role for industrial robots where robots do not substitute humans, but instead they work side by side with them [12]. Based on this, one can think on scenarios where robots and humans could share the same tools, without the need to buy and place expensive additional mechanisms for the robot every time we need the robot to handle specific objects. To the best of the author's knowledge, there have not been works presented in the literature focused on this kind of tools.

Fig. 2. Commercial grippers. On the left, The SRDK-UR from Universal Robots. It imitates the human hands in terms of flexibility, dexterity, and precision [11]. On the right, the FlexShape Gripper by Festo which can grip different types of objects regardless their shape.

The design and development of the tool is part of the PIR (Perception for Industrial Robots) project from the Pontificia Universidad Javeriana Bogota (PUJ) [6]. This project is focused on characterizing and simulating the application of an industrial robot for waste separation tasks and conducting research on the technical aspects of the collaborative and flexibility concepts of COBOTs. In this article, we propose a vacuum tool that can be used for picking plastics bottles in the bottle classification task conducted in the Alquería recycling center in Bogotá. There, tasks are conducted manually, thus this being a perfect scenario for testing the collaboration between humans and robots (see Fig. 3).

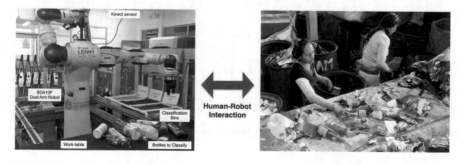

Fig. 3. Human-Robot Interaction proposed for waste classification tasks at the Alquería recycling center [10].

The Alquería recycling center is responsible for separating color High Density Polyethylene (HDPS), High Density Polyethylene/Polystyrene (HDPS/PS), green polyethylene terephthalate (PET), and transparent PET, among others. Due to the wide variety of bottles that arrive to the recycling center, e.g. different sizes; the robot grippers, in some cases, are not the appropriate tools for grasping the containers. Therefore, within the project we propose a tool magazine that will be in the robot's workspace and will be accessible either by the robot or the operator, if required. In the magazine, different vacuum tools of different sizes can be available.

The paper is organized as follows, Sect. 2 shows the designed criteria considered for designing the tool. Section 3 presents the prototype. Section 4 describes the tests carry out to evaluate the tool, and Sect. 5 presents the conclusion and the direction of future work.

2   Design Criteria

Grippers are tools that are only designed for robots. However, the proposed tool targets both robots and humans.

To meet the human's specifications, anthropometric measurements of the hand of adults (shown in Fig. 4) and ergonomic parameters that are based on the type of activity to be developed, were considered. Table 1 summarizes the parameters considered when designing the tool from the human point of view, taking into account the application (pick and place of plastic bottles).

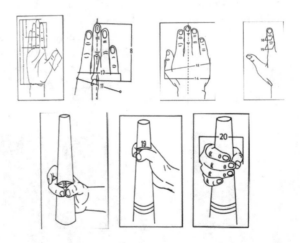

Fig. 4. Anthropometric measures of the hand of an adult for holding object [8].

Additional aspects were considered in the design of the tool:

- The tool should encourage the use of both hands, to help mitigating the problems of left-handed people and fatigue when the dominant hand is fatigued. For more than 90% of users, the dominant hand is the right one.
- The design should follow strictly the design criteria, this because an intense compression of the tool leads to the inflammation of the tendon sheaths, compromising blood supply and leading to an excessive compression of critical nerves [14].
- The weight of the tool should not exceed 3 Kg. This to comply with the limitations stablished by the ISO 11228 about manual handling [3].

Table 1. Ergonomic parameters considered for the design of the tool.

Factor	Parameters
Activity	**Type of task**. Pick and place
	Work environment. Workers are standing up in an 8-h shift. The operator must pick and place the bottles in a basket that is close to the table
Postural and biomechanical analysis	**Body posture**. Standing when handling plastic bottles
	Hand posture and hand-wrist movement. Natural posture of the wrist
The tool	**Shape of the tool**. The handle must provide maximum contact between the tool and the skin (or glove). In general, its transverse section should be cylindrical, flat, or elliptical [7]
	Dimension and weight. The user must be able to use the tool with one hand. Ideal weights of tools operated by the hand-forearm segment is 0.9–1.5 Kg. Length of the handle for force tools: 100–125 mm
	Surface of the handle. Should ensure a good grip on a handle. There must be enough friction between the hand and the handle. Handles made of plastic or composite rubber are recommended

In addition to the human factor, the robot's gripping system was considered when designing the tool. In this study, the hardware shown in Fig. 5 was used. It is comprised of the UR3 robot from Universal robots, and the two fingers gripper from Robotiq. The parameters of the robot that are considered for the design of the tool are shown in Table 2. In addition to those criteria, the tool

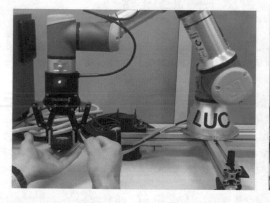
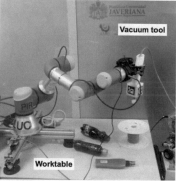

Fig. 5. Testbed. It is comprised of the UR3 collaborative robot from Universal robots equipped with the two fingers gripper from Robotiq.

must be easily identified, so that the robot can automatically locate the tool when it needs it.

Table 2. Robot parameters

Characteristics	Value
Robot payload	3 kg
Friction grip payload	3 kg
Gripper weight	1 kg
Stroke	50 mm
Grip force	60 to 130 N

3   Prototype

Figure 6 summarizes the criteria considered for designing the tool. As It is shown in the image, the tool was designed considering some criteria only for the human (body posture, hand posture, etc.); but additionally there are other criteria such as, shape, weight, and surface, that required to be evaluated for both the human and the robot.

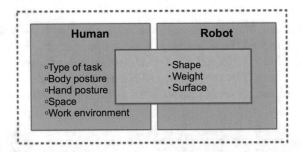

Fig. 6. Ergonomic and robot parameters for the design of the tool. Shape, Weight, and Surface criteria should comply the design criteria for both, the human and the robot.

Figure 7 shows the designed tool. The upper part (marked with number 1), correspond to the handle, where the operator and the robot will hold the tool. The lower part (marked with number 2) corresponds to a cube, where the suction cup will be placed in. The lower part was designed to allow the placement of visual markers on the tool (ArUco codes) for its automatic detection and identification.

The handle of the tool was designed with a length of 123 mm and a diameter of 38 mm. In addition, the handle has an inclination of 12°, following ergonomics recommendations for manual tools. The tool will have also a button to allow manual activation of the vacuum.

Fig. 7. Prototype of the tool. The upper part (number 1), correspond to the handle that will be shared by the robot and the operator. The lower part (number 2) will contain the succion cup and visual markers for automatic detection and identification of the tool.

4 Test and Results

Three different tests were conducted to analyze the design of the tool. The first one was focused on the robot; the second one, analyzing people's opinion about the tool; and the third one corresponds to a test where the robot and a person shared the tool.

4.1 Robot

In this test, the UR3 robot was used to automatically detect the tool and pick up 9 bottles of different sizes grouped as small, medium, and big (three bottles per size). This test was focused on verifying that the robot can pick up the tool and that it does not slip from the gripper's fingers (i.e. that the bottle does not fall). Figure 8 shows the robot and some images of the bottles the robot had to pick up.

The test consisted on commanding the robot to pick up the bottles. For each group of bottles (small, medium, and big), the test was conducted 10 times at 3 different speeds, low (10% mx. robot's speed), medium 10 (50% mx. robot's speed) and high (90% mx. robot's speed). The maximum speed of the UR3 robot is 1 m/s. If the bottles of each group were successfully picked up, the test was considered successful, otherwise failed.

The results of the test are summarized in Table 3. From this table, it can be seen that the system picked up all the small bottles and that the change of speed did not affect the results. On the other hand, the robot succeeded in 93% of the tests, when picking up the medium size bottles; and 83% of the tests, when picking up the big size bottles. Analyzing the results, the 7 trials when

Bottles to Pick

Fig. 8. Setup Test 1. The robot shown on the left image was commanded to pick up 3 bottles of each group 10 times at three different speeds (30 trials).

the robot failed carrying the bottles correspond to tests conducted at high speed and when the vacuum tool was far from the center of gravity of the bottle.

Table 3. Results Test 1. Robot picking up 3 bottles of each group 10 times at three different speeds (30 trials). If all the bottles of each group were successfully picked up, the test was considered successful, otherwise failed.

Bottle's size	Successful trials	Failed trials
Small	30	0
Medium	28	2
Big	25	5

4.2   Ergonomic Test

A test was carried out to analyze the people's opinion regarding the tool. Five persons where questioned about ergonomic and manipulation aspects of the proposed tool when using the tool for picking up the bottles. Each person handled 10 plastic bottles with the tool and answered the following survey. The answer for each question was limited to Good, Regular, or Bad.

1. Is the handle length enough for the size of your hand?
2. Is the handle diameter comfortable?
3. Is the weight of the tool considered to be adequate?
4. Is the position of the suction cup correct?
5. Is the position of the activation button of the suction cup correct?

Table 4 summarizes the results obtained in the survey. It shows the number of people whose answer coincided in each question. The following list summarizes the people's comments for each question.

1. **Handle length.** People commented that the length of the tool was appropriate for their hands and they were able to manipulate the tool without problems. They recommended to improve the material of the handle to prevent it form slipping.
2. **Handle diameter.** People with small hands recommended to reduce the diameter of the handle to make it more comfortable for them.
3. **Tool's weight.** Everybody agreed that the weight was appropriate for manipulating the tool.
4. **Suction cup.** Most of them commented that it was difficult to see the suction cup.
5. **Activation button.** They agreed that the position of the activation button was correct, but that it required a lot of force to be activated.

Table 4. Results Test 2. The table summarizes the number of people whose answer coincided for each of the question related to ergonomic and manipulation aspects, when using the tool.

Question	Scores		
	Good	Regular	Bad
1	4	1	0
2	3	2	0
3	5	0	0
4	1	4	0
5	2	3	0

4.3   Human-Robot Test

A final test was conducted to have both, the human and the robot using the tool for picking up plastic bottles. Figure 9 shows representative images from the test, where the human is manipulating the tool for picking up some bottles.

Figure 10 shows images when the robot was conducting the task. The tool was automatically detected and identified by a computer vision algorithm. ArUco markers located on the four sides of the cube were used to provide fast 3D pose estimations of the tool to the robot. With this markers, the human can place the tool in any orientation and the system can still detect the tool.

A video of the test can be seen in [5].

Fig. 9. Results Test 3 Human Manipulating the Tool. Representative images of the test when the human uses the tool for picking up some bottles.

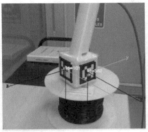

Fig. 10. Results Test 3 Robot Manipulating the Tool. Image on the left shows the computer vision algorithm the robot uses to detect and identify the tool. The other two images are representative images of the test when the robot picks up some bottles with the tool.

5   Conclusions

In this paper we presented the design of a vacuum tool that can be shared by humans and robots, for plastic classification tasks. Real tests were conducted, and the results showed that the robot was able to successfully pick up of small bottles at different speed. With the medium size and big size bottles, some of the bottles fell off when operating at high speed. However, it is important to mention that the weight of those bottles was greater than the weight the suction cup can handle.

From the results, we can say that the vacuum tool can be carry by the robot appropriately, since in none of the tests the tool fell off the robot. Future work is focused on improving ergonomic aspects of the tool, based on the suggestions provided by the people involved in the tests.

Acknowledgements. This work has been supported by the Pontificia Universidad Javeriana Bogota. Project 7697, PIR Perception for Industrial Robots. The authors would like to thank the CTAI for the support with the equipments.

References

1. AER: Robotica colaborativa. https://www.aer-automation.com/mercados-emerge ntes/robotica-colaborativa/
2. ATI-Industrial-Automation: ATI tool changer products: MC-10 (2018). https:// www.ati-ia.com/products/toolchanger/QC.aspx?ID=MC-10
3. Becker, J.P.: Las normas iso 11228 en el manejo manual de cargas (2009). http:// www.semac.org.mx/archivos/congreso11/Pres09.pdf
4. Crowe, S.: 6 innovative robotic grippers lend a helping hand (2018). https://www. therobotreport.com/6-innovative-robotic-grippers-lend-helping-hand/
5. CTAI: A collaborative vacuum tool for humans and robots (2018). https://www. youtube.com/watch?v=5tzehMjjQJQ
6. CTAI: Pir recycling (2018). https://www.youtube.com/watch?v=F76Pe-WkP3g& t=2s
7. Ergonomia108: Diseño y seleccion de herramientas (2013). http://ergonomia108. blogspot.com/2013/10/23-diseno-y-seleccion-de-herramientas.html
8. Jain, S., Pathmanathan, G.: Importance of anthropometry for designing user-friendly devices: mobile phones. J. Ergon. (2012). https://doi.org/10.4172/2165-7556.1000109
9. Kerak, P., Holubek, R.: Automatic gripper exchange in intelligent manufacturing systems, pp. 1313–1314 (2011)
10. Lizarazo, N.B., Rico, D.G., Martinez, C., Atuesta, S.B.: Designing a framework to give perception capabilities to an industrial robot for waste separation tasks, p. 39 (2017)
11. Universal Robots: Efectores finales: Cobots con destreza humana (2018). https:// blog.universal-robots.com/es/efectores-finales-cobots
12. Universal Robots: La industria 4.0: Camino hacia la automatización y la van-guardia (2018). https://blog.universal-robots.com/es/industria-40
13. Schlichtb, L.B.T.: A statistical review of industrial robotic grippers, pp. 88–97 (2018). https://doi.org/10.1016/j.rcim.2017.05.007
14. de Seguridad e Higiene en el Trabajo (INSHT): Herramientas manuales. crite-rios ergonomicos y de seguridad para su seleccion (2016). https://www.insst.es/ InshtWeb/Contenidos/Documentacion/

Designing an Interface for Trajectory Programming in Industrial Robots Using Augmented Reality

Juan C. Gallo and Pedro F. Cárdenas[✉]

Department of Mechanical and Mechatronic Engineering,
National University of Colombia, Bogotá, Colombia
{jucgallopi,pfcardenash}@unal.edu.co

Abstract. Traditional programming methods of industrial robots, show different disadvantages such as the use of highly qualified personnel and long programming time for the generation of robot trajectories. The development of friendly and intuitive user interfaces that improve the human-robot interface is greatly important; so it allows generating trajectories in a natural way. To achieve this goal, tools such as augmented reality are required to enhance the robot's environment with virtual information in which the user may receive feedback. It is proposed to develop an immersive user interface using augmented reality, that allows programming trajectories for the ABB IRB 140 robot by means of a cubic marker and giving useful feedback to the user throughout the device THC VIVE. The system of augmented reality is developed with Vuforia in Unity3D, and the system of communication in ROS.

Keywords: Augmented reality · Robot programming · Human-robot interface

1 Introduction

Several industrial sectors are currently facing a globalized and highly competitive world, with short life cycles for their products and an increasingly demanding market. This requires the incorporation of industrial robots to make the processes more efficient; so this requires natural and friendly interfaces that: encourage human-robot interaction, increase maneuverability, and reduce user fatigue, during programming, manipulation and control tasks. Therefore, it is necessary to provide ways for robots and humans to interact among them in a natural, intuitive and unequivocal way.

Traditionally robot programming systems can be classified into two large methods [11]. The first one is *programación online* that encompasses the *Lead-through* and *Walk-through* techniques, which use teach pendant devices and force sensors respectively in order to save specific positions on the controller, and then by using them to create motion sequences. *programación online* is not intuitive

A. Martínez et al. (Eds.): LACAR 2019, LNNS 112, pp. 142–148, 2020.
https://doi.org/10.1007/978-3-030-40309-6_14

and has several disadvantages, such as the need for highly qualified programmers and the use of long programming time, and verification that become dead production time.

The second method is the *programación offline* (OLP) that allows the robot to be programmed remotely in a virtual environment that simulates the operating conditions of the robotic cell. This requires 3D models of the robot and the entire environment around it; great effort, high programming time, and abundant investment in simulation software. In spite of *programación offline* does not incur in dead production time, *programación online* is more widely accepted and used.

With the aim of correcting these shortcomings and encouraging the use of industrial robots, friendly interfaces are required, but also easy and quick of programming, and at the same time providing safety and feedback to the operator. This program should enable natural interaction with the robot without prior knowledge of robot programming [4]. Augmented reality (AR) is a tool of human-machine interaction that overlays virtual information on the real scene. In this way, AR can be combined with human skills to aid manufacturing tasks [8]. Some studies have shown the advantages this technique has in industrial sectors, such as: reduced time, errors and mental effort, improving spatial capacity, and accelerating learning [5].

This is the reason why AR is a powerful tool to improve productivity, so its use in industrial applications has been evident. Some of these apps have been developed in areas such as: maintenance, assembly and repair, training, quality control, construction monitoring and facility monitors [9]. The use of this technique has not been alien to robotic systems; so that the *robot programming using augmented reality* (RPAR) has become a new paradigm of production.

The system developed by Fang et al. [2,3], presents an RPAR interface, whose objective is to assist the user in programming a virtual robot in a real environment, for the planning of Pick and Place tasks or tracking path. The programming is carried out with the help of a cubic marker, which replaces the teach pendant and allows to define the points of operation and the orientation of the end effector.

The AR enhances reality and improves situational awareness, because it provides the user with information on the processes of assembly, visualization of trajectories, visual alerts, and production data for the case of lines of automotive assembly [6] and [7], or information such as the speed of a conveyor belt, the current tool equipped for the CNC machine and the coordinates of the manipulated object, for the case of a robotic cell [10].

The app presented by Akan et al. [1], shows an AR interface that highlights the workpieces as green virtual objects, the user can select which parts to move and where to move them; the AR interface will show the selected part in red and the path that the robot will run in yellow. On the other hand, Quintero et al. [12] proposes an AR system that highlights the robot's trajectory, previews the robot's movements and displays parameters through AR glasses.

2   Methodology

The proposed system for programming trajectories in an industrial robot using RPAR is based on the recognition and tracking of high contrast markers. Key points are extracted from these markers which are then stored in the application. A camera captures the robot's scene and each frame performs the recognition process. If there is a marker in the scene, the application calculates the position and orientation of the marker with respect to the camera and renders the virtual objects inside the captured scene. See Fig. 1A.

A simple marker called UN is used, as shown in Fig. 1B with its specific key points and a cubic marker as shown in Fig. 1C, which is an array of simple markers that set key points in 3D. After that, a UN marker is attached to a framework that is the origin of the whole scene O_s and the center of the cubic marker that replaces the teach pendant, the framework of the O_t programming tool is attached. A probe is associated with the cubic marker to facilitate to the user the selection of points that will make part of the robot's trajectory; and the O_p framework is attached at the end of the probe.

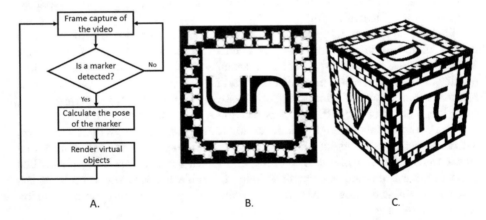

Fig. 1. A. Recognition and Tracking Flow B. Simple UN marker and key points C. Cubic Marker

The camera recognizes within the scene the marker UN and the cubic marker, and the pose of each of them is calculated with respect to the origin of the camera O_c, and with this, the transforms of the origin O_t to O_c are defined and called $^c_t T$ and the transforms of the origin O_s to O_c are defined and called $^c_s T$. The O_r framework is attached to the origin of the robot and the transformation from the O_s to O_r origin is $^r_s T$.

The UN marker stays fixed in a known position; regarding this, the pose of each point in the path is calculated and saved. The marker cubic will be mobile in order to make it easier for the user to program the points of the trajectory; and thanks to its 3D configuration, it will enable its recognition from different points of view. To select a point in ^pP space, the cubic marker probe is located near the point of interest; and then, the camera recognizes the cubic marker, and calculates its position and orientation, with relation to the origin of the O_c camera. See Fig. 2. The calculation of each ^pP point concerning O_s will be:

$$^sP = {}^c_sT^{-1}\ {}^c_tT\ {}^t_pT\ {}^pP \tag{1}$$

The calculation of ^sP regarding the robot's origin O_r is:

$$^rP = {}^r_sT\ {}^sP \tag{2}$$

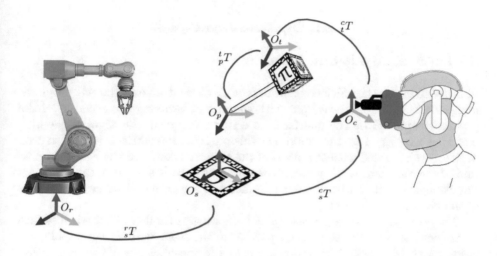

Fig. 2. Transformaciones del sistema

The system uses a web camera that captures the image of the scene and allows the markers' recognition and tracking. A Head Mounted Display (HMD) device for this HTC VIVE enables visual feedback to the user of the probe position and orientation attached to the cubic marker, as well as the robot's trajectory and alert messages. Subsequently the path edited by the user utilizing AR is executed by the ABB manipulator IRB-140.

The general architecture of the proposed system is shown in Fig. 3. The user interface is developed in Unity3D that renders virtual objects regarding markers that are recognized with the Vuforia Augmented Reality SDK. The position and orientation of the points edited with the cubic marker probe are extracted from the scene and sent to ROS through ROS-Sharp and executed by ROS-Industrial.

Figure 4 shows a screenshot of what the user can see through the AR interface. The system allows rendering the frameworks of the UN marker and the cubic marker, getting information such as alerts or the pose of the programming tool, as well as the path that the robot executes when finishing the at points' edition. The path to be followed by the robot is shown with yellow lines that join the points, and the orientation that the robot must get at each point is shown with orange 3D rectangle.

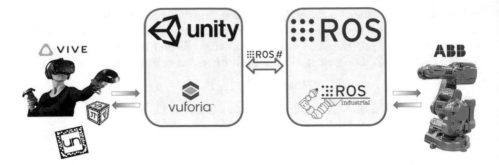

Fig. 3. General architecture system

3   Data Analysis and Results

In order to check the efficiency and reliability of the tool and the AR programming interface, a study was done with the aim of verifying the accuracy of the positions captured by the interface. To achieve this goal, the edges of the blue part shown in Fig. 4 are measured. For this purpose, the vertices pointed to each of them are recorded with the probe of the cubic marker; and the Euclidean distance between them is calculated. The data obtained is shown in Table 1 where the distance attained with the AR interface and the actual distance of the edges of the piece are compared.

The error rate observed in the Table 1, shows great ability of the AR interface to extract the points of the robot's path from the scene with an acceptable precision for certain tasks. This error rate is due to the efficiency of the recognition and tracking algorithms, even so it should be noted that much of this error is the result of the visual estimation of the user and the error caused by the sensitivity of touch.

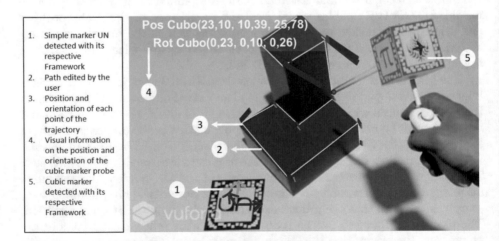

Fig. 4. View of the user's interface

Table 1. Medidas capturadas con la interfaz AR

Edge	AR measurement (mm)	Actual measurement (mm)	Error %
1	85	88	3.6
2	192	198.3	3.3
3	171	176.2	3.0
4	93	88.7	4.6
5	85	80.7	5.1
6	100	99.2	0.8
7	192	191	0.5
8	100	99.4	0.6
9	85	79	7.1
10	100	94.5	5.5
11	85	82.9	2.5

4   Conclusions

Robot programming using Augmented Reality can represent a paradigm shift in traditional programming methods. The RPAR enables to carry out programming tasks of the robot much faster, naturally and without requiring prior knowledge, which makes it a promising tool to improve the human-robot interaction.

The measurements made with the AR interface demonstrated high reliability, keeping the error at a very low percentage; and though so far the accuracy of the recognition and tracking systems are not ideal for tasks that require great precision such as welding and assembly jobs, the RPAR can be very useful for Pick and Place tasks where the tolerance margin is much higher.

Such kinds of systems are quite vulnerable to the conditions of illumination, recognition problems generated by the distance between the camera and the markers, or the speed with which the camera and/or the cubic marker moves. These problems can be overcome to the extent that the recognition algorithms are more efficient.

It could also be noticed that RPAR demands the user's large concentration in order to keep the UN marker inside the scene and locate the cubic marker so that it is identified in the desired orientation. This is done while the probe tip is placed in the position of interest to achieve the selection of the points of the trajectory. Therefore, it is recommended as future work to add a set of simple markers to the system configuration that work as a reference, so that if a marker is occluded or lost in the scene, there will be another that easily replaces it.

References

1. Akan, B., Ameri, A., Cürüklü, B., Asplund, L.: Intuitive industrial robot programming through incremental multimodal language and augmented reality. In: Proceedings - IEEE International Conference on Robotics and Automation, pp. 3934–3939 (2011). ISBN 9781612843865
2. Fang, H.C., Ong, S.K., Nee, A.Y.: A novel augmented reality-based interface for robot path planning. Int. J. Interact. Des. Manuf. **8**(1), 33–42 (2014). ISSN 19552513
3. Fang, H.C., Ong, S.K., Nee, A.Y.: Novel AR-based interface for human-robot interaction and visualization. Adv. Manuf. **2**(4), 275–288 (2014). ISSN 21953597
4. Ioanes, C., Chioreanu, A.: Current trends regarding the intuitive programming of industrial robots. Acta Technica Napocensis **55**(I), 207–210 (2012)
5. Jetter, J., Eimecke, J., Rese, A.: Augmented reality tools for industrial applications: what are potential key performance indicators and who benefits? Comput. Hum. Behav. **87**(February), 18–33 (2018). ISSN 07475632
6. Makris, S., Karagiannis, P., Koukas, S., Matthaiakis, A.S.: Augmented reality system for operator support in human-robot collaborative assembly. CIRP Ann. Manuf. Technol. **65**(1), 61–64 (2016). ISBN 0007–8506
7. Michalos, G., Karagiannis, P., Makris, S., Tokçalar, Ö., Chryssolouris, G.: Augmented reality (AR) applications for supporting human-robot interactive cooperation. Procedia CIRP **41**, 370–375 (2016). ISBN 2212–8271
8. Nee, A.Y., Ong, S.K.: Virtual and augmented reality applications in manufacturing. IFAC Proc. Vol. (IFAC-PapersOnline) **46**, 15–26 (2013). ISBN 9783902823359
9. Pace, F.D., Manuri, F., Sanna, A.: Augmented reality in industry 4.0. Am. J. Comput. Sci. Inf. Technol. **6**(1), 1–7 (2018). ISSN 23493917
10. Pai, Y.S., Yap, H.J., Md. Dawal, S.Z., Ramesh, S., Phoon, S.Y.: Virtual planning, control, and machining for a modular-based automated factory operation in an augmented reality environment. Sci. Rep. **6**(February), 1–19 (2016). ISSN 20452322
11. Pan, Z., Polden, J., Larkin, N., Van Duin, S., Norrish, J.: Recent progress on programming methods for industrial robots. In: Joint 41st International Symposium on Robotics and 6th German Conference on Robotics, ISR/ROBOTIK 2010, vol. 1, pp. 619–626 (2010). ISBN 9781617387197
12. Quintero, C.P., Li, S., Pan, M.K., Chan, W.P., Van Der Loos, H.M., Croft, E.: Robot programming through augmented trajectories in augmented reality. In: IEEE International Conference on Intelligent Robots and Systems, pp. 1838–1844 (2018). ISBN 9781538680940

Modeling and Antibalance Control
of a Birail Crane

Jessica Garzón, Jenny Alfonso, Liliana Fernandez-Samacá,
and Camilo Sanabria[⊠]

UPTC, Tunja, Colombia
{jessica.garzon01,jennymarcela.alfonso,
liliana.fernandez,camilo.sanabria}@uptc.edu.co

Abstract. This paper presents the modeling of a didactic birail crane, and the design of a controller by state space to control the oscillation of the load when the operator displaces it. For the design of the controller, an observer of complete order states was considered, which allows estimating the values of the states that cannot be measured and filtering the measurable state variables. The modeling of the system was performed using Euler-LaGrange, taken as state variables, the position of the trolley, the angular position of the pendulum (oscillating load) and the respective velocities. The didactic prototype has a range of 2 m in the horizontal and 1 m in the vertical axis and uses DC motors as actuators. The control of the prototype is made by using an application (App) designed in Android Studio. The controller is synthesized in a DsPic 30F4013 microcontroller, which facilitates the updating of the controller and implementation of new control laws.

1 Introduction

The birail cranes are one of the systems most used in the industry to transport loads in large places like warehouses, marine ports and airports. There are different types of birail cranes which are adjusting according to the needs, for example, distances and weight [1,3].

One of the main challenges of this systems is to control the oscillations that may occur when transporting the load, to avoid losses of materials and time, or accidents. Therefore, to design a controller to minimize load's oscillations and facilitate the operator work results is an interesting study case in industrial control. This work focuses on the design and construction of a didactic prototype of a birail crane, which seeks to emulate its dynamic behavior.

For controlling the pendulum oscillation, authors designed a state space feedback controller, which uses a state observer to estimate non-measurable variables and filter the measurable ones [5].

There are many contributions about control techniques for the birail crane, like PID based on the transfer function of the system [5], state space feedback [2], Kalman filters [6], fuzzy logic control designs of the Takagi-sugeno type [4], and

© The Editor(s) (if applicable) and The Author(s), under exclusive license to
Springer Nature Switzerland AG 2020
A. Martínez et al. (Eds.): LACAR 2019, LNNS 112, pp. 149–156, 2020.
https://doi.org/10.1007/978-3-030-40309-6_15

sliding modes by using DSP [7], among others. The prototype presented herein seeks to serve as a platform for testing different control techniques; for this reason, authors have included a DsPic microcontroller to synthetizes different control algorithms; likewise, the used sensors and motors are easily accessible.

The rest of the document is organized as follows, Sect. 2 presents the modeling of the system, Sect. 3 describes the design of the prototype and Sect. 4 shows the design of the controller and the state observer. Finally, Sect. 5 presents the analysis of the results and Sect. 5 concludes the work developed.

2   System Modeling

The birail crane system has two degrees of freedom; one is the movement of on the horizontal axis trolley (position x), and another is the load oscillation (θ, the angle between the pendulum and vertical axis), see Fig. 1. The mathematical modeling of the birail crane is made by using the Euler - LaGrange formulation, based on the kinetic and potential energy produced by the movement. The system model is expressed in state space, that take as state variables x, θ, v and ω. Where v and ω are the linear and angular velocity.

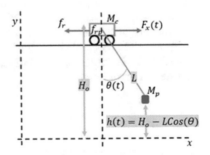

Fig. 1. Force diagram.

Where: M_c: Mass of the car.
M_p: Mass of the pendulum.
$x(t)$: Trolley position.
$x'(t)$: Trolley speed.
$\theta(t)$: Angular position of the pendulum.
$\theta'(t)$: Angular velocity of the pendulum.
f_r: Trolley friction.
f_{rp}: Pendulum friction.
F_x: Force applied by the engine to the car.

From analysis of the kinetic and potential energies for both the trolley and the pendulum, the authors calculate the Lagrangian formulation for obtaining the non-linear state space equations, which is shown in the Eqs. (1) and (2).

$$
\begin{bmatrix}
X_3 \\
X_4 \\
-\frac{f_{rp}X_3}{L^2 I(X_1)}\left(\frac{M_c}{M_p}+1\right)+\frac{f_r X_4 \cos X_1}{IX_1 L}-\frac{M_p \sin 2X_1 X_3^2}{2IX_1}-\frac{(M_c+M_p)g\sin X_1}{LIX_1} \\
-\frac{f_r X_4}{IX_1}+\frac{M_p L \sin X_1 X_3^2}{IX_1}+\frac{\cos X_1 f_{rp} X_3}{IX_1 L}+\frac{M_p g \sin 2X_1}{2IX_1}
\end{bmatrix}
=
\begin{bmatrix}
F_1(x) \\
F_2(x) \\
F_3(x) \\
F_4(x)
\end{bmatrix}
\tag{1}
$$

$$
G(x) =
\begin{bmatrix}
0 \\
0 \\
-\frac{\cos X_1}{ILX_1} \\
\frac{1}{I(X_1)}
\end{bmatrix}
\tag{2}
$$

Later, the own values of the system were calculated and, the obtained model was linearized around to an equilibrium point, $theta = 0$, $v = 0$ and $\omega = 0$, obtaining the model expressed in (3).

$$
X'(t) = AX(t)Bu(t)
$$
$$
y(t) = CX(t)
\tag{3}
$$

where:

$$
A =
\begin{bmatrix}
0 & 0 & 1 & 0 \\
0 & 0 & 0 & 1 \\
\frac{-(M_c+M_p)g}{LM_c} & 0 & \frac{-f_{rp}}{L^2 M_c}\left(\frac{M_C}{M_P}+1\right) & \frac{f_r}{LM_c} \\
\frac{M_p g}{M_c} & 0 & \frac{f_{rp}}{LM_c} & -\frac{f_r}{M_c}
\end{bmatrix}
; B =
\begin{bmatrix}
0 \\
0 \\
\frac{1}{M_c} \\
\frac{1}{M_c}
\end{bmatrix}
; C = \begin{bmatrix} 1 & 0 & 0 & 0 \end{bmatrix}
$$

Equation (4) express the final model for the birail crane prototype, taking into account its parameters shown in the Table 1.

$$
A =
\begin{bmatrix}
0 & 0 & 1 & 0 \\
0 & 0 & 0 & 1 \\
-37.31 & 0 & -0.768 & 1.822 \\
19.678 & 0 & 0.405 & -1.44
\end{bmatrix}
; B =
\begin{bmatrix}
0 \\
0 \\
-8 \\
8
\end{bmatrix}
\tag{4}
$$

3   Design and Implementation of the Prototype

The design of a birail crane prototype was made by using low-cost and materials available in the local market. The prototype has a box in which the battery and circuits are stored; three motors (M1, M2, M3), M1 moves the trolley in a metallic structure that has two rails embedded, M2 put up and down the load and

Table 1. System parameters

Parameter	Value
M_c	0.125 Kg
M_P	0.251 Kg
g	9.8 m/s^2
L	0.79 m
f_r	0.18
f_{rp}	0.04

M3 that receive the control signal. For measuring the behavior of variables, the prototype has two linear potentiometers as position sensors (Sp), two snap-action switches (Fc), and an ultrasonic sensor (Sd) to measure the angular position (θ). The structure of the prototype is implemented using iron profiles, one pipe of 1 in2 (square-inch), in which the rails for the trolley are Fig. 2 shows the location of sensors and motors.

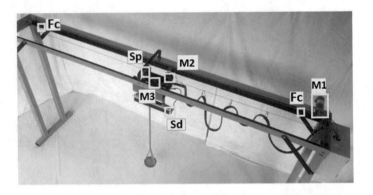

Fig. 2. Metallic structure of the birail crane prototype.

Figure 3 shows the trolley, and the designed interface, the trolley it is made in acrylic and it is anchored by screws and industrial glue, Trolley has a plate for putting the load. For the system of trolley displacement on rails were included bearings. The operator of the birail crane prototype can handle it using an application made in Android Studio where its interface with buttons allows to command the horizontal and vertical displacement of the load.

App interface manual control crane.   **System for horizontal displacement of trolley on the birail crane**

Fig. 3. Trolley on the birail crane prototype and App interface for user

4   Controller and State Observer Design

The control challenge for the system is to reduce the load oscillations while it is transported. For which, authors design a controller by State Space Feedback.

The controllability analysis of the discrete system is performed (Eq. (5)) to determinate the rank of the matrix of Eq. (6).

$$x(k+1) = G(k)x(k) + H(k)u(k) y(k) = C(k)x(k) + D(k)u(k) \qquad (5)$$

Where:

$$G = \begin{bmatrix} 0.9926 & 0 & 0.0198 & 0.0004 \\ 0.0039 & 1 & 0.0001 & 0.0197 \\ -0.7317 & 0 & 0.9776 & 0.0356 \\ 0.384 & 0 & 0.0118 & 0.9718 \end{bmatrix} ; H = \begin{bmatrix} -0.0016 \\ 0.0016 \\ -0.1555 \\ 0.1569 \end{bmatrix} ; C = \begin{bmatrix} 1 & 0 & 0 & 0 \\ 0 & 1 & 0 & 0 \\ 0 & 0 & 1 & 0 \\ 0 & 0 & 0 & 1 \end{bmatrix} ; D = 0$$

$$M = \begin{bmatrix} H & G & ... & G^{n-1}H \end{bmatrix} \qquad (6)$$

The rank of the matrix M is 4 verifying that the system is totally controllable. To determine the constants of the controller (Eq. (7)), the authors used the pole assignment method in Matlab® the system and controller are tested by using Simulink.

$$u(k) = -Kx(k) \qquad (7)$$

The determined constants are:

$$K = \begin{bmatrix} -31.4406 & 10.2183 & 3.8176 & 7.994 \end{bmatrix}$$

For designing a space state feedback control system, it is necessary to know all state variables, but in actual cases, not all variable can be measured directly; therefore, the designer must estimate those non-measured variables [5]. For this estimation, an observer of complete order states is designed, which also allows filtering the noises that appear in the measurements.

The observability analysis verifies the minimum conditions of observability, the rank of Matrix O (Eq. (8)) is 4 and corresponds to the size of the matrix of states.

$$O = \begin{bmatrix} C \\ CG \\ \vdots \\ CG^{n-1} \end{bmatrix} \tag{8}$$

The prediction observer is defined by Eq. (9), where Ke is the gain matrix of the observer, which is determined from Eq. (10), by using the duality principle.

$$x(k+1) = (G - K_eC)x(k) + Hu(k) + K_ey(k) \tag{9}$$

$$K_e = K^* \tag{10}$$

By replacing the respective values of the known matrices, the feedback gain matrix of the observer is that shown in ((11)).

$$K = \begin{bmatrix} -31.440 \\ 10.2183 \\ 3.8176 \\ 7.994 \end{bmatrix} \tag{11}$$

For the observer implementation of the in the microcontroller, the code for the controller is changed, calculating the non-measurable variables. For synthesizing the controller, a DsPic 30f4013 microcontroller is used.

5   Analysis and Results

Figure 4 show the results obtained for the controller and observer of states designed for the prototype. The four state variables are observed. When a disturbance of 15° is applied to θ, the controller acts rejecting the disturbance, the output signal shows that θ value goes to the equilibrium point (0°) the controller responds in the time for which it was designed, the results obtained with the implementation of the observer of states shows that: the control objective is met (eliminating the roll of the load), the movements made by the car (position in x) are smoother and lastly that the control signal is less demanding with respect to that required in the implementation of the controller. Proving that the objectives are fully met, as regards the rejection of disturbances in the oscillation of the load and the response time of the system.

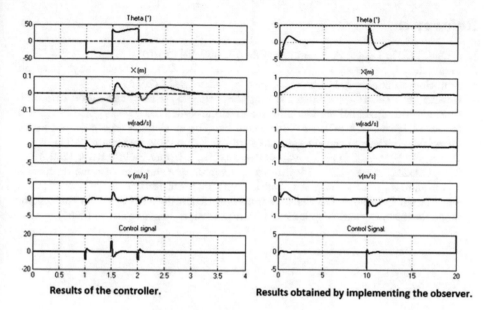

Results of the controller.   Results obtained by implementing the observer.

Fig. 4. Results obtained.

6   Conclusions

Obtaining the model from the Lagrange analysis, for this type of systems, is one of the most successful, given that it is easier to obtain state space analysis, thus allowing each of the variables involved in the system.

The design of the prototype is based on materials, sensors and actuators of easy access, which makes it a very economical and easy assembly prototype, for the synthesis of the controller a microcontroller is used which allows the implementation of multiple control techniques, making it a useful tool for teaching.

To evaluate the performance of the designed prototype, the technique of controller feedback of states is implemented, without including the dynamics of load elevation, which would be one of the future works on the educational prototype; As a result, the stabilization of the load at the desired equilibrium point is obtained, in addition, a state observer is implemented as a filter of the measurable variables and prediction of those that are not, obtaining very good results in terms of the response of the same. The algorithms were implemented using the compiler MPLAB and synthesized in a DsPIC 30f4013.

Given that the prototype was built for educational purposes, as future work, the implementation of different control techniques is considered, in order to improve the behavior regarding the rejection of the disturbances that occur in the pendulum that holds the load, including filters digital data for the collection of data from the sensors, in addition to the implementation of different status observer typologies to verify their functionality.

References

1. Collado, R.E.: Diseno de puente grúa de 5 toneladas (2010)
2. Fernández, L., Castro, I., Mora, J., Salamanca, J.: Control "anti sway" para un prototipo de puente grúa birraíl.L'esprit Ingénieux **4**(1) (2015)
3. Gordón, M., Cicerón, S., Silva Proaño, C.S.: Diseño de un puente grúa tipo de 5 toneladas de capacidad para la industria metalmecánica. B.S. thesis, QUITO (2011)
4. Navarro Chávez, G.A.: Diseño e implementación de un sistema de control por lógica difusa del tipo takagi-sugeno aplicado a un prototipo de grúa-puente (2011)
5. Ogata, K.: System Dynamics, vol. 4. Prentice Hall, Upper Saddle River (2004)
6. Ríos, C.D.Z., Suárez, E.G.: Identificación robusta aplicada a un sistema de control de un puente grúa. Scientia et technica **18**(3), 437–446 (2013)
7. San Martín Castillo, P.D.: Diseño e implementación de un sistema de control por modo deslizante usando dsp aplicado a un prototipo de grúa puente (2005)

Sensorless Control of an Induction Motor with Common Source Multilevel Converter

Edison Andrés Caicedo Peñaranda, Jorge Luis Díaz Rodríguez[✉],
and Luis David Pabón Fernández

Universidad de Pamplona, Ciudadela Universitaria, Pamplona, Colombia
jdiazcu@gmail.com

Abstract. This work deals with the development of sensorless control for the induction motor by means of a three-phase multilevel converter with optimization of the harmonic content in line voltages. The control was designed and implemented, based on the frequency variation method implemented in an H-bridge multi-level converter of common source in the range of 5 Hz to 100 Hz. This involves the development of a multiobjective optimization algorithm (switching angles, THD and voltage level), the sensorless control technique feeds back the control loop with the estimated speed of the current magnitudes based on the mathematical model of the induction motor. The sensor system is embedded in the converter eliminating the need to use mechanically coupled sensors in the induction motor, the control technique works properly in steady and dynamic mode, and the harmonic content (THD) is optimized being less than 2%.

Keywords: Induction motor · Multilevel converter · Sensorless control · THD · Genetic algorithm

1 Introduction

Direct and indirect conversion power converters are widely used in modern industry [1], such as frequency converters, applied in the speed control of induction motors [2]. In these converters, by varying the frequency of the motor supply, the mechanical speed of the motor can be controlled. This process is carried out by generating voltages of variable frequency and magnitude [3, 4]. A classic speed control method is the application of the scalar control technique or simply V/Hz technique. The variation in this ratio is used with the objective of not saturating the electrical machines. The supply RMS voltage must decrease as the frequency decreases from nominal value [5]. In general, conventional converters using pulse width modulations (PWM) are used to perform this change in voltage magnitude and frequency. These modulations vary the frequency and magnitude of the voltage by controlling the duration of the pulses within the waveform and the period of the pulse [6].

Power converters have problems of power quality [7], because their output voltages have a square pulse waveform, which causes a high harmonic content [8]. Leading to the windings heating and the generation of opposite parasitic torques in the induction machines [7, 9]. The voltage waveform depends on the voltage level required and the

© The Editor(s) (if applicable) and The Author(s), under exclusive license to
Springer Nature Switzerland AG 2020
A. Martínez et al. (Eds.): LACAR 2019, LNNS 112, pp. 157–171, 2020.
https://doi.org/10.1007/978-3-030-40309-6_16

frequency. The THD value of the converter changes as the frequency and the RMS value of the voltage varies, becoming a function of the RMS value and the switching frequencies THD (Vrms, f) [10]. In order to solve this problem of harmonic distortion in the DC/AC conversion of inverters, multilevel power converters have appeared as an attractive alternative [11, 12]. Several works are achieved in the area of the optimization of the harmonic content of the inverters at constant frequency [13–15] and some approximations in the application [16, 17], also of the optimization in the area of the variable frequency in the control of electric machines [17–19]. Ratifying the benefits that multilevel power inverters have for powering electric machines [21, 22], mainly because their use implies advantages in terms of power quality, energy efficiency, equipment conservation. Achieving that the control of induction machines develops complex refinement methodologies [23–27].

The multilevel inverter topology of cascaded H-bridge with common source according to some previous works is not recommended for applications of variable frequency due to the saturation of the transformers [28]. However, it can be evidenced that using an appropriate transformer design methodology [29] the converter can be used in these applications, an so on as frequency inverter for the control of induction motors under the sensorless control technique [30–33].

2   The Multilevel Converter and Its Control

The multilevel power inverter topology implemented is presented in Fig. 1. This corresponds to a common source asymmetric cascaded H-bridge converter with a transformers ratio of 1:3 and two H-bridges stages each phase, providing 9 levels of phase voltages and 15 levels of line voltages [13, 34].

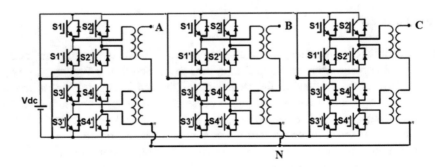

Fig. 1. Multilevel power converter topology.

The algorithm that allows to vary the frequency and magnitude the voltage with the lowest possible THD uses the line voltage and the THD value in terms of the switching angles per phase. For which the Fourier series of the phase voltages with symmetry of

1/4 of the waveform were determined. Allowing to create the modulation with the switching angles in the first quarter of the waveform. The other parts of the phase voltage modulation and the other two phases are constructed by trigonometric relationships.

In this way we define a vector $L = [a\ b\ c\ d]$ that represents the total number of on and off angles in each level (at level 1, b level 2, c level 3 and level 4) and with the difference of the phases determines those of the line voltages.

The Fourier series for the phase A voltage is defined by:

$$h_n = \begin{cases} 0 & \text{for } n \text{ even} \\ \frac{4v_{cd}}{\pi n} \sum_{i=1}^{4} \sum_{j=1}^{L_i} (-1)^{j-1} \cos(n\alpha_{ij}) & \text{for } n \text{ odd} \end{cases} \tag{1}$$

Where v_{dc} is the voltage value of each step, and α_{ij} is the $j(a,b,c,d)$ angle of the $i(1\ 2\ 3\ 4)$ step. For phase B, which is shifted $120°$ respecting the phase A, the Fourier series will be defined by:

$$h_n = \begin{cases} 0 & \text{for } n \text{ even} \\ \frac{4V_{dc}}{\pi n} \sum_{i=1}^{4} \sum_{j=1}^{L_i} (-1)^{j-1} \cos n\alpha_{ij} & \text{for } n \text{ odd multiple of three} \\ \frac{4V_{dc}}{\pi n} \sum_{i=1}^{4} \sum_{j=1}^{L_i} (-1)^{j-1} \cos n\alpha_{ij} & \text{for } n \text{ odd not multiple of three} \end{cases} \tag{2}$$

Performing the differences (in terms of the Fourier series) of voltages in the phases A and B. The Fourier series for the line voltage v_{AB} is obtained by:

$$h_n = \begin{cases} 0 & \text{for } n \text{ even} \\ 0 & \text{for } n \text{ odd multiple of three} \\ \frac{4\sqrt{3}V_{dc}}{\pi n} \sum_{i=1}^{4} \sum_{j=1}^{L_i} (-1)^{j-1} \cos n\alpha_{ij} & \text{for } n \text{ odd not multiple of three} \end{cases} \tag{3}$$

2.1 Line Voltages and THD

The IEEE 519 defines the total harmonic distortion (THD) up to the 50th harmonic [35]. Where the h_1 harmonic is the fundamental component and h_n is the peak of the harmonic n:

$$THD = \frac{\sqrt{\sum_{n=2}^{50} \left\{ \frac{1}{n} \left[\sum_{i=1}^{4} \sum_{j=1}^{L_i} (-1)^{j-1} \cos n\alpha_{ij} \right] \right\}^2}}{\left[\sum_{i=1}^{4} \sum_{j=1}^{L_i} (-1)^{j-1} \cos 1\alpha_{ij} \right]} \cdot 100 \tag{4}$$

Where n takes odd values with harmonics multiples different from three. Because the harmonics multiples of three are suppressed in the line connection of the transformers. That is to say 5, 7, 11, 13, 17,... and L_i are the components of the vector $L = [a\ b\ c\ d]$.

Similarly, the RMS value can be defined in terms of switching angles and harmonics according to Eq. (5):

$$V_{line_{RMS}} = \sqrt{\sum_{n=1}^{50} \frac{\left\{ \frac{4\sqrt{3}V_{CD}}{\pi n} \left[\sum_{i=1}^{4} \sum_{j=1}^{L_i} (-1)^{j-1} \cos n\alpha_{ij} \right] \right\}^2}{2}} \tag{5}$$

$n = 5, 7, \cdots$, odd multiples different of three

2.2 Algorithm for the Determination of Line Voltages

To avoid the saturation of electrical machines, the multilevel power converter reduces the line voltage in terms proportional to the frequency [6]. This is done by means of a scalar technique obtained with an algorithm for the search of a modulation with a determined RMS value by means of Eq. (5) as a restriction equation and with the minimum of THD with Eq. (4) as objective (*fitness*) function to be minimized. For the development of the search algorithm of the modulations in the whole frequency range, it uses a genetic algorithm [36] which is presented in Fig. 2.

In the algorithm as the first step, the rated values of the motor are assigned, such as the rated line voltage (V_n) and the rated frequency (f_n). The frequency range over which the control technique is assigned ($[f_i, f_f]$) is also entered and the frequency step with which the voltage/frequency ratio (Δf) is calculated. Finally, the maximum number of generations with which the multiobjective genetic algorithm (N_{max}) should operate in each modulation is set. With this data the algorithm calculates a vector with the frequencies and a vector with the voltages.

2.3 Sensorless Control

The developed control is based on the principle of the loop shown in Fig. 3. Where the value of the rotor speed depends on an estimator based on the proportionality of the rotor speed and the imaginary current of the space vector on the rotor axis in the Stationary reference frame using the Clarke and Park transformations, which determines the slip used to calculate the rotor speed using the synchronism speed [4].

The controller was programmed using Labview software as suggested by the Clarke and Park equations [4]. The estimator is multiplied by a gain, which is determined by the open-loop operation of the entire system. The rotor speed is connected to the controller input as presented in Fig. 4.

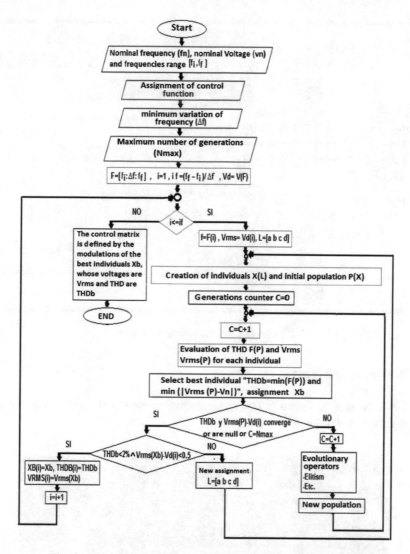

Fig. 2. Flowchart of the used multiobjective genetic algorithm.

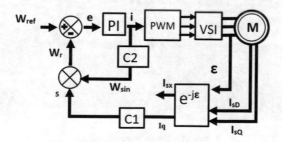

Fig. 3. Implemented control loops.

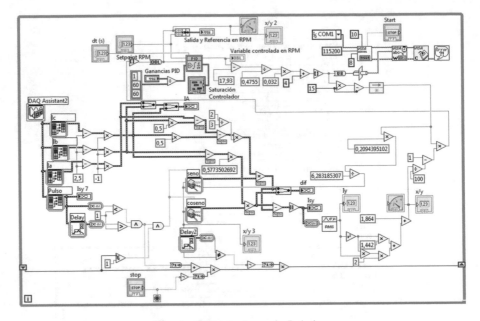

Fig. 4. Control scheme in Labview.

3 Performance Tests

The converter used is shown in Fig. 5, according to the Fig. 1 scheme and the transformer design methodology [29].

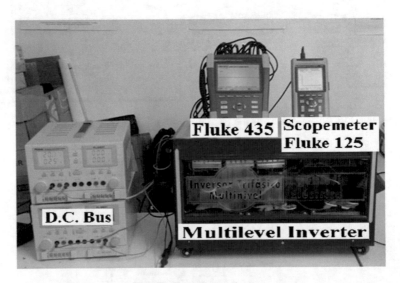

Fig. 5. Experimental setup.

3.1   Line Voltages and THD Value

The induction motor used operates at a rated frequency of 50 Hz. For which the designed frequency range is [1 Hz, 100 Hz] with steps of $\Delta f = 0.5$ Hz, the V_{boost} level of 30 Vrms of line is obtained 200 modulations with low harmonic content and with RMS values that follow a V/Hz control technique. Table 1 shows a fragment of the switching angles obtained.

Table 1. Optimized switching angles base on the best individual

f	k	a	b	c	d	TC	V_{RMS}	THD	Angles		
1	1	121	0	0	0	121	33.8	0.2742	0.07759	0.16753	2.9178
1.5	1	121	0	0	0	121	35.7	0.0754	1.02185	1.03020	2.9183
2	1	85	0	0	0	85	37.6	0.0394	2.52739	2.55259	4.3227
2.5	1	89	0	0	0	89	39.5	0.0686	1.64416	1.65871	3.9459
3	1	85	0	0	0	85	41.4	0.1286	1.13806	1.16112	4.4966

The frequency of each modulation is shown in the first column of the table. In the second column the value of k is shown, which is the number of steps that the modulation uses in the first quarter of the waveform without counting the zero step, a, b, c and d define the vector L. In the table only uses "a" due to the low voltage level required for these frequencies. TC is the total number of commutation angles, V_{RMS} is the root mean square voltage obtained and the THD is the total harmonic distortion of the modulation.

Figure 4 shows the graph of the control technique V/Hz with the RMS values determined by the algorithm.

Figure 5 shows the THD values versus frequency. The results show that in most frequencies they are very close to 0%. However, in the range of [40.5 42 Hz] the THD value exceeds 1% and presents a maximum peak of 1.8% (Figs. 6 and 7).

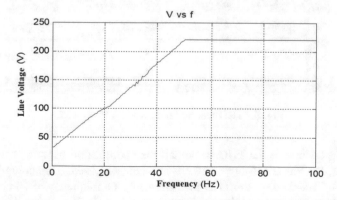

Fig. 6. V/Hz ratio determined by the genetic multiobjetive algorithm.

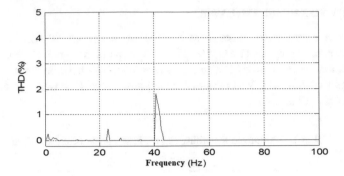

Fig. 7. Chart of the THD values vs frecuency.

3.2 Line Voltages and THD

The measurements of the line voltage and the THD values obtained with the Fluke 125 type oscilloscope are presented in Fig. 8. Where the operation of the converter at 6 different time instants and with frequencies below and above the rated (50 Hz) is observed with reduced THD. Verifying the proper functioning of the proposed algorithm.

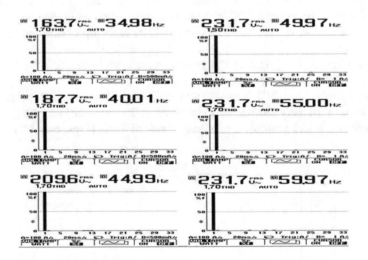

Fig. 8. THD measurement using a Fluke 125.

For the validation of the THD value and the optimization process at the values of line voltages, the Fluke 434 type network analyzer was used. This complies with the IEC 61000-4-7 standard for the THD measurement, at a frequency of 60 Hz. Figure 9a shows the phase voltages and their harmonic spectrum. In Fig. 9b the THDv value of 10.9% due to the presence of multiples of three harmonics which are suppressed in the line voltage waveforms as shown in Fig. 10a. Its harmonic spectrum is shown in

Fig. 10b which shows a measured THD value of 0.6%. Verifying the reduction of the THD value of the line voltages proposed in the algorithm and evidenced differences with the Fluke 125 oscilloscope in the measured THD value.

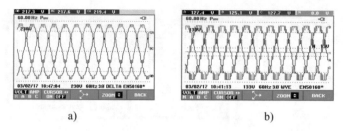

a)   b)

Fig. 9. Voltages waveform at 60 Hz (a) phase (b) line.

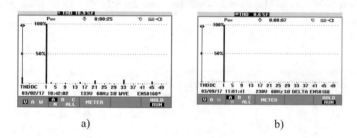

a)   b)

Fig. 10. Harmonic spectrum at 60 Hz (a) phase (b) line.

3.3 Controller Design

The proposed controller was tuned with the help of Matlab software. The transfer function was obtained from the identification of the open loop system. Considering that the computer sends the information of the modulations to the FPGA that controls the multilevel converter. This at the same time is connected to the three-phase induction motor, where the current sensors are installed and send the information to the computer, processing the speed as indicated in Fig. 11.

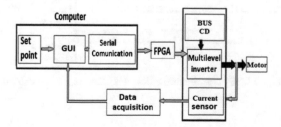

Fig. 11. Block diagram of the system identification.

The identification process was carried out using the "*ident*" toolbox of Matlab. The data obtained from the system of Fig. 11 before a motor speed step and the results of the model are presented in Fig. 12.

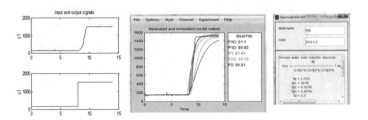

Fig. 12. Matlab's system identificacion process.

The characteristics of the tuned controller and the simulation in Matlab are shown in Fig. 13.

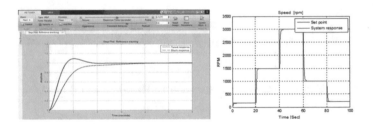

Fig. 13. Controller tuning.

The general scheme of the controller implemented in a block diagram is presented in Fig. 14.

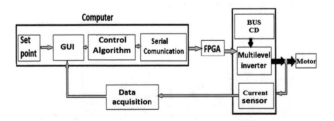

Fig. 14. System identification with the control algorithm.

3.4   Sensorless Control Validation

Figure 15 presents the capture of the graphical interface against three changes of reference speed, the first of 1000 rpm, the second of 1500 rpm and the third is 700 rpm. It can be seen in the graphical interface the changes were assigned to the controller (green), with the values corresponding to the previous calculation showing the proper functioning of the feedback with the speed estimator.

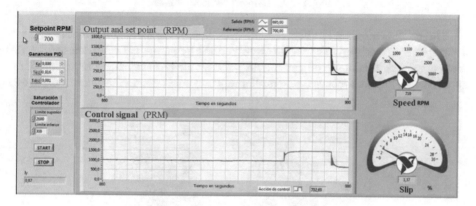

Fig. 15. Labview control interface.

To verify the proper functioning of the controller in the event of a disturbance, the torque and speed of the induction motor was measured in response to an increase in torque as shown in Fig. 16, at an operating point of 1600 rpm.

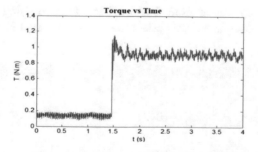

Fig. 16. Torque response.

The magnitudes of the load in terms of power measured by the network analyzer are presented in Fig. 17.

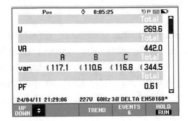

Fig. 17. Electric power measurement.

The result is presented in Fig. 18. The operation of the induction motor at speed of 1600 rpm is appreciated. The values were verified by an optical speed measuring element (encoder). It is observed that after 8 s the torque rises causing a reduction of the speed, before this disturbance the controller regulates and restores the value of the speed.

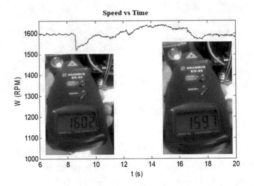

Fig. 18. Speed verification.

It is evidenced by testing from the visual interface verifying that the controller responds to disturbances in an appropriate manner as indicated in Fig. 19.

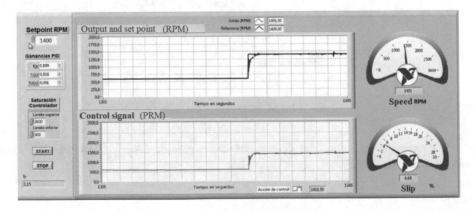

Fig. 19. Controller response with disturbance.

The induction motor was loaded with a DC generator with resistive electrical load. Being this abruptly switched causing an instantaneous change in the output torque of the motor that reduces the speed and raises the rotor currents. This causes the increase of the stator currents that are measured and transformed to correct the speed.

4 Conclusions

The developed sensorless controller responds appropriately to changes in load rise and load fall. Additionally, the tests with and without disturbances evidence the adequate integration of the improvement of the THD in the develop V/Hz control technique. In spite of the 0.5 Hz frequency steps, the controller has the capacity to assume different speed magnitudes to the common ones by combining modulations automatically. Where speed estimation provides the appropriate indications to achieve this operation. The system is limited in the response due to data transmission and processing in milliseconds.

The V/Hz control technique found in this work allows the frequency to be varied from 1 Hz to 100 Hz, proportionally changing the voltage as the frequency decreases from its rated value. However, the algorithm can take any form of equation that involves the terms V and f. Because the proposed optimization is performed directly on line voltage, multiple harmonics of three can exist in the phases. Because these are eliminated by realizing the potential difference between them, therefore, in the line voltages will not exist. This is verified in the developed mathematical modeling.

The induction motor control system when using a multilevel power converter and harmonic optimization, offers advantages such as the reduction of voltage and current harmonics in accordance with that presented in Fig. 10. This theoretically decreases the ripple in the electromagnetic torque that has an appreciable effect on the output torque as shown in Fig. 16, directly affecting the behavior of the motor speed and currents.

References

1. Abu-Rub, H., Holtz, J., Rodriguez, J., Baoming, G.: Medium-voltage multilevel converters —state of the art, challenges, and requirements in industrial applications. IEEE Trans. Ind. Electron. **57**(8), 2581–2596 (2010)
2. Leon, J.I., Kouro, S., Franquelo, L.G., Rodriguez, J., Wu, B.: The essential role and the continuous evolution of modulation techniques for voltage-source inverters in the past, present, and future power electronics. IEEE Trans. Ind. Electron. **63**(5), 2688–2701 (2016)
3. Fitzgerald, A.E., Kingsley, C., Umans, S.D.: Electric Machinery, 6th edn. McGraw Hill, New York (2003)
4. Vas, P.: Sensorless Vector and Direct Torque Control. Clarendon Press, Oxford (1998)
5. Rodriguez, J.L.D., Fernandez, L.D.P., Peñaranda, E.A.C.: Multiobjective genetic algorithm to minimize the THD in cascaded multilevel converters with V/F control. Appl. Comput. Sci. Eng. 742, 456–468 (2017)
6. Chapman, S.J.: Máquinas eléctricas, 5th edn. McGraw-Hill Interamericana, New York (2012)

7. Sánchez, M.A.: Calidad de la energía eléctrica. Instituto Tecnológico de Puebla, México (2009)
8. Yumurtaci, M., Ustun, S.V., Nese, S.V., Cimen, H.: Comparison of output current harmonics of voltage source inverter used different PWM control techniques. WSEAS Trans. Power Syst. **3**, 696–703 (2008)
9. Barbera, G., Mayer, H.G., Issouribehere, F.: Medición de la emisión armónica en variadores de velocidad y desarrollo de modelos de simulación. Encuentro Regional Iberoamericano de CIGRE (2009)
10. Fernandez, L.D.P., Rodriguez, J.L.D., Arevalo, A.E.: Multilevel power converter with variable frequency and low constant total harmonic distortion. In: IEEE 5th Colombian Workshop on Circuits and Systems (CWCAS), Bogotá, (2014)
11. Malinowski, M., Gopakumar, K., Rodriguez, J., Pérez, M.: A survey on cascaded multilevel inverters. IEEE Trans. Ind. Elect. **57**(7), 2197–2206 (2010)
12. Kavali, J., Mittal, A.: Analysis of various control schemes for minimal total harmonic distortion in cascaded H-bridge multilevel inverter. J. Electr. Syst. Inf. Technol. **3**, 428–441 (2016)
13. Rodriguez, J.L.D., Fernandez, J.L.P., Garcia, A.P.: THD improvement of a PWM cascade multilevel power inverters using genetic algorithms as optimization method. WSEAS Trans. Power Syst. **10**, 46–54 (2015)
14. Babu, T.S., Priya, K., Maheswaran, D., Kumar, K.S., Rajasekar, N.: Selective voltage harmonic elimination in PWM inverter using bacterial foraging algorithm. Swarm Evol. Comput. **20**, 74–81 (2015)
15. Rodriguez, J.L.D., Fernández, L.D.P., Peñaranda, E.A.C.: A genetic optimized cascade multilevel converter for power analysis. In: Communications in Computer and Information Science, vol. 657, pp. 123–137 (2016)
16. Chitra, A., Himavathi, S.: Reduced switch multilevel inverter for performance enhancement of induction motor drive with intelligent rotor resistance estimator. IET Power Electron. **8**, 2444–2453 (2015)
17. Taleba, R., Benyoucefa, D., Helaimia, M., Boudje, Z.: Cascaded H-bridge asymmetrical seven-level inverter using THIPWM for high power induction motor. Energy Procedia. **74**, 844–853 (2015)
18. Letha, S.S., Thakur, T., Kumar, J.: Harmonic elimination of a photo-voltaic based cascaded H-bridge multilevel inverter using PSO (particle swarm optimization) for induction motor drive. Energy **107**, 335–346 (2016)
19. Shriwastava, R., Daigavaneb, M.B., Daigavanec, P.M.: Simulation analysis of three level diode clamped multilevel inverter fed PMSM drive using carrier based space vector pulse width modulation (CB-SVPWM). Procedia Comput. Sci. **79**, 616–623 (2016)
20. Manasa, S.: Design and simulation of three phase five level and seven level inverter fed induction motor drive with two cascaded h-bridge configuration. Int. J. Electr. Electron. Eng. **1**, 2231–5284 (2012)
21. Fernández, L.D.P., Rodríguez, J.L.D., García, A.P.: Simulación del inversor multinivel de fuente común como variador de frecuencia para motores de inducción. Revista de Investigación, Desarrollo e Innovación **7**(1), 1–10 (2016)
22. Antonopoulos, A., Mörée, G., Soulard, J., Ängquist, L., Nee, H.: Experimental evaluation of the impact of harmonics on induction motors fed by modular multilevel converters. In: Electrical Machines (ICEM), Berlin (2014)
23. Ghosh, E., Mollaeian, A., Hu, W., Kar, N.C.: A novel control strategy for online harmonic compensation in parametrically unbalanced induction motor. IEEE Trans. Magn. **52**(7), 1–4 (2016)

24. Tatte, Y.N., Aware, M.V.: Direct torque control of five-phase induction motor with common-mode voltage and current harmonics reduction. IEEE Trans. Power Electron. **32** (11), 8644–8654 (2017)

25. Mahato, B., Raushan, R., Jana, K.C.: Modulation and control of multilevel inverter for an open-end winding induction motor with constant voltage levels and harmonics. IET Power Electron. **10**(1), 71–79 (2016)

26. Tatte, Y.N., Aware, M.V.: Torque ripple and harmonic current reduction in a three-level inverter-fed direct-torque- controlled five-phase induction motor. IEEE Trans. Ind. Electron. **64**(7), 5265–5275 (2017)

27. Ghosh, E., Mollaeian, A., Kim, S., Tjong, J., Kar, N.C.: DNN-based predictive magnetic flux reference for harmonic compensation control in magnetically unbalanced induction motor. IEEE Trans. Magn. **53**(11), 1–7 (2017)

28. Panda, A.K., Suresh, Y.: Research on cascade multilevel inverter with single DC source by using three-phase transformers. Int. J. Electr. Power Energy Syst. **40**, 9–20 (2012)

29. Diaz-Rodriguez, J.L., Pabon-Fernandez, L.D., Caicedo-Peñaranda, E.A.: Novel methodology for the calculation of transformers in power multilevel converters. Ingeniería y competitividad **17**, 121–132 (2015)

30. Verma, V., Chakraborty, C., Maiti, S., Hori, Y.: Speed sensorless vector controlled induction motor drive using single current sensor. IEEE Trans. Energy Convers. **28**(4), 938–950 (2013)

31. Zhang, X.: Sensorless induction motor drive using indirect vector controller and sliding-mode observer for electric vehicles. IEEE Trans. Veh. Technol. **62**(7), 3010–3018 (2013)

32. Kumar, K., Chauhan, Y.K., Shrivastava, V.: A fuzzy logic based sensorless induction motor drive supplied from a photovoltaic source. In: 2013 International Conference on Microelectronics, Communications and Renewable Energy (AICERA/ICMiCR), Kanjirapally (2013)

33. Yang, H., Zhang, Y., Walker, P.D., Liang, J., Zhang, N., Xia, B.: Speed sensorless model predictive current control with ability to start a free running induction motor. IET Electr. Power Appl. **11**(5), 893–901 (2017)

34. Fernández, L.D.P., Rodríguez, J.L.D., García, A.P.: Total harmonic distortion optimization of the line voltage in single source cascaded multilevel converter. WSEAS Trans. Syst. **15**, 199–209 (2016)

35. IEEE: IEEE Std. 519-1992 IEEE Recommended Practices and Requirements for Harmonic Control in Electrical Power Systems. IEEE (1992)

36. Goldberg, D.: Genetic Algorithms in Search, Optimization and Machine Learning. Addison-Wesley, Boston (1989)

Application of the Watershed Segmentation Method in the Separation and Identification of Individual Leaves in Potato Crops

Hernán Javier Guio Carrillo[✉] and Adriana Lucia Villamizar Fuentes

Universidad de Pamplona, Pamplona, Colombia
hernan0ll4@gmail.com,
adriana.villamizar2@unipamplona.edu.co

Abstract. This research aims to find and to identify early symptoms of the presence of pests in potato plants crops through the application of techniques and algorithms of artificial vision and classification of their characteristics. For this, the Watershed algorithm is used in order to segment and to identify the largest number of possible individual leaves present in a conventional agricultural field and then, to make a classification using a neural network to determine, by means of the characteristics of the morphological structure, the leaves that have any type of noticeable infection for their early and efficiently control.

This Process is developed by using technological tools such as the implementation of an unmanned aircraft that would allow large aerial video shots, helping producers to reduce time and money; and to improve the quality of the food by containing less amount of chemicals that are harmful to the people's health.

Keywords: Segmentation · Watershed · Artificial vision · Pest detection

1 Introduction

Usually, potato production is affected by different types of pests that could damage the quality of the orchards, affecting growers' economy. Most farmers do not have enough staff and time to verify, day by day, their crops conditions, leaving the potatoes plants vulnerable to the latent threat of pests.

Regarding this problem, the idea of the computer vision techniques application emerged by design a software that allows farmers to determine automatically, by giving the characteristics and environmental conditions of the leaves, if they suffer from pest symptoms such as the presence of holes caused by Thysanoptera and Saltonas flies. By implementing this software, the efficiency of detection and prevention of potatoes crops pest could be improve. Besides, by combining it with other tools as the use of drones that favor aerial imagery, it could be possible to cover large areas developing a quality evaluation of the crops in less time and decreasing costs for producers.

A. Martínez et al. (Eds.): LACAR 2019, LNNS 112, pp. 172–184, 2020.
https://doi.org/10.1007/978-3-030-40309-6_17

2 Detection Method

The steps of the execution of the algorithm are the following:

2.1 Background Removal

In a complex field environment, there are several factors that affect the segmentation of the target leaves. For example, the presence of soil, stems, flowers, stones and water pipes. Therefore, it is necessary to select only the areas of interest (leaves) and ignore the regions classified as background. For this, the levels of the green channel are used in the RGB color space [1].

According to the principle of color gradient, the ratio of the color channels of pixels for a green crop leave is shown in (Eq. 1)

$$(G - B > \theta_1) \cap (G - R > \theta_2) \tag{1}$$

Where G, B and R represent the values of the green, blue and red color channels respectively and θ_1 and θ_2 represent control parameters that were established as $\theta_1 = 10$ and $\theta_2 = 15$ (Fig. 1).

Fig. 1. Background removal

2.2 Segmentation by Using the Watershed Algorithm

To understand the Watersehd algorithm operation, a grayscale image should be displayed as a 3D image. In this case the image is seen as a topographic surface in which the high intensities resemble peaks and hills, while low intensities appear as valleys. You should suppose, the farmer start to pour water of different colors, indicating labels on each isolated valley, which corresponds to a minimum present. When the water level begins to rise, depending on the nearby peaks, the water of different valleys with different colors will begin to merge. To avoid this, morphological techniques are used to guide the algorithm to areas with foreground objects to be segmented (in this case, leaves), and barriers are built in the places where the water is joined. Then, the work of filling with water and building barriers continues until all the peaks are under water. Thus, the barriers created in this process are the segmentation of the image.

2.2.1   Steps to Guide Segmentation

The magnitude of the gradient is used as the segmentation function, for this the Sobel mask is used. It calculates the intensity of the gradient, it is high in the borders of the objects and lowers (especially) in its interior (Fig. 2).

Fig. 2. Magnitude of the gradient

Then, the objects of the foreground are marked by morphological techniques such as opening by reconstruction and closure by reconstruction. These operations will create maximum planes within each object (Fig. 3).

Fig. 3. Close-up objects marked

Next, you must calculate background markers. In the clean image, the dark pixels belong to a second plane that is not of interest in this case, the background (Fig. 4).

Fig. 4. Background marked

Once it is done, the Watershed transform is calculated having the foreground objects and the marked background (Fig. 5).

Fig. 5. Result of the segmentation

2.3 Classification of Segments

After detecting the segments by the algorithm it is important to try to determine which of these leaves are truly, you could use the morphological information and extract key features that help you to identify target leaves. These pickpockets could be:

2.3.1   Relationship Between Axes

The horizontal and vertical size of each leave is calculated, followed by a numerical relationship between both measurements.

By analyzing these values, it was determined that the ratio varies from 1.1 to 1.8 in different samples; this value shows the change in the proportion of one axis with respect to the other (Eq. 2), therefore, those segments that do not are included in this range will not be taken into account.

$$Relationship_{Axis} = \frac{Higher\ axis}{Minor\ axis} \qquad (2)$$

2.3.2   Perimeter of the Image

In the next step, we proceed to traverse and mark the pixels around the image boundary, these values are assigned in an array as coordinates points (x, y) (Fig. 6).

Fig. 6. Leaf perimeter

2.3.3   Centroid

We obtain a horizontal coordinate x, a vertical coordinate and the center of mass of the extracted leaf segment. Once the data is available, the distance from the centroid of each image to each point found in the perimeter is calculated (Fig. 7).

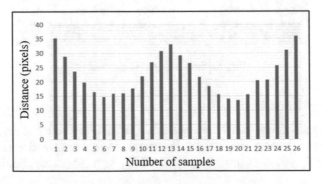

Fig. 7. Distance from the center to the perimeter of a true *sheet*

According to the curve that is interpreted in the previous results, it is possible to distinguish that a true leaf is soft while but, a false leaf has a noisier behavior. Knowing this, you could use a statistical parameter as the standard deviation to see how varied or dispersed these values are. The different images could have differences in the magnitudes of their distances because of the resolution, it is necessary to compare this data with information referring to the same leave.

Through the experimentation of several appropriates and inappropriate leaves, it was determined that the leaves of interest are below a ratio of 30 (Eq. 3) with respect to the estimated standard deviation and average.

$$\frac{(Standard\ deviation * 100)}{Average\ Distances} \tag{3}$$

2.3.4   Segment Area
You get the real number of pixels in the region that includes the leave returned as scalar. This data is useful to discard those areas that are very large or very small. To implement this information in the procedure, the resolution of the image that is being worked is taken into account, it directly influences the results.

2.3.5   Convex Envelope Area
This technique specifies the coating or coverage that encloses all pixels within the filled target leave.

Once the normal area and the area of the convex envelope of the image have been obtained, the relationship of (Eq. 4) is used to determine those leaves that are in a better state since, those imperfect or damaged edges leaves will present a major convex envelope area (Fig. 8).

Fig. 8. Convex envelope area of a true and false leaf

$$Relationship\,Areas = \frac{Normal\,Area}{Convex\,Enviromental\,Area} \tag{4}$$

2.3.6   Fourier Analysis

When dealing with the distances calculated from the center to the edges we will obtain values that can be interpreted as an analog signal. Taking that it will be possible to apply the Fourier transform to appreciate the noise that exists in the leave by examining the spectrum of frequency distribution (Fig. 9).

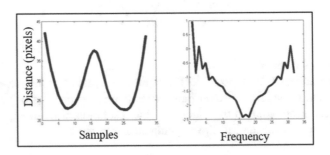

Fig. 9. Distance and frequencies of a true leaf

According to the figure, it could be seen the greater oscillation in the frequencies, this is because the initial signal that indicates the distances is more variable since the values are not homogeneous.

2.3.7   Neural Network Training

Backpropagation is a method of gradient calculation which is used in supervised learning algorithms as the simple perceptor model with logistic function, it consists of a training set of input-output pairs where you will find a function that given a new input infers the expected output. This model is useful for predicting the presence or absence of a characteristic or result; based on the values of a set of predictor variables. The advantage of the logistic regression is the possibility of including in the model continuous or discrete variables and any combination of them without the need to present a normal distribution [2].

The input set that will be used to perform the training of the neural network with logistic function will be the data obtained with the relation of the standard deviation between the center and the perimeter, the relationship between the axes (x, y), the relationship between the normal area against the area of the convex envelope of the segments and the Fourier analysis. These values will be the input of the network. Moreover, in the output there will be values of 1, for the segments that are leaves and 0 for those segments that are not.

It was collected 100 data from different segments, 50 belonging to leaves and 50 that do not represent segments of interest. By using these values, the network will be trained by using the cross validation technique used to evaluate the results of a statistical analysis and guarantee which are independent of the partition between training and test data.

This technique consists of repeating and calculating the arithmetic mean obtained from the evaluation measures on different partitions. In this case, there will be 10 different groups; where 90 data will be the training data and the remaining 10 will be the test data. In this way, the network will have different values to verify the learning. With this network we try to identify all those segments that are not leaves in order to discard them in the next stage of the process since they could interfere in the final result.

2.4 Hole Detection

Once the objects are identified and marked as leaves, we proceed to examine if they have any type of orifice produced by any kind of pest by identifying spaces with a great difference in color within the segmented area. The same process is carried out on each segment (Fig. 10).

Fig. 10. Identification of each segment in the image

3   Results and Analysis

To test the algorithm, 5 images of potato crops were taken with a Drone outside of the Pamplona. To check the efficiency of the algorithm, the number of leaves in the foreground was manually counting in each image and then it was compared with the results of the proposed method (Fig. 11).

The total of resulting segments thrown by the algorithm was classified through the implementation of the neural network. The following table shows the tests obtained from the success in separating leaves according to the algorithm by means of the cross validation method (Table 1).

Fig. 11. Potatoes crop image taken with a Drone

Table 1. Cross validation neural network

Test number	Segments that are not leaves (0)					Segments that are leaves (1)					% Succes
1	1	1	0	0	0	1	1	1	1	1	80
2	0	0	0	0	0	1	1	1	1	1	100
3	0	0	0	0	1	1	1	0	1	1	80
4	0	0	0	0	1	1	1	1	1	1	90
5	0	0	0	0	0	1	1	1	1	1	100
6	0	0	0	0	0	1	1	1	1	1	100
7	0	0	0	0	0	1	1	1	1	1	100
8	1	0	1	1	0	1	1	1	1	1	70
9	0	0	0	0	0	1	1	1	1	1	100
10	0	0	0	0	0	1	1	1	1	1	100
Average											92

There is a success rate of 92%, this means that the training of the neural network obtained a high percentage of success, classifying most of the segments correctly.

It was possible to eliminate a significant amount of elements that did not represent importance when increasing efficiency by 95% (Table 2).

Table 2. Segments classification

Image of potato planting	Total unsorted segments	Segments classified as leaves	Improvement percentage
1	756	43	94,3
2	806	44	94,5
3	2947	84	97,1
4	883	40	95,5
5	891	44	95,1
Average	1256,6	51	95,3

The results are categorized into two main sections, the true segments (real leaves found) and false segments (abstract segments that are not leaves) (Table 3).

Table 3. Leaves category identified by the algorithm

Image of potato planting	Segments found by the algorithm	S egments that are leaves	Segments that are not leaves
1	43	41	2
2	44	42	2
3	84	74	10
4	40	38	2
5	44	36	8
Average	51	46,2	4,8

Having the number of segments thrown by the algorithm and the number of real leaves could be found the success of discarding (Eq. 5).

$$\frac{Classified\ leaves * 100\%}{True\ leaves} \tag{5}$$

9% of the segments launched by the algorithm are not individual leaves, but represent multiple occluded leaves or parts of divided leaves.

While 91% represent the leaves of interest, these would be the appropriate ones to examine the health of the crop and determine its status (Fig. 12).

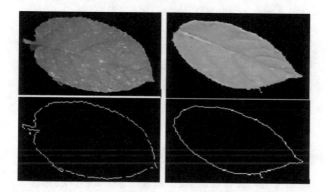

Fig. 12. Individual segmented leaves

Now, to determine the success of finding good leaves against the number of foreground leaves counted manually it is determinate as follows (Eq. 6).

$$\frac{Classified\,leaves\,*\,100\%}{Sheets\,counted\,manually} \tag{6}$$

The algorithm finds 8% of the actual leaves in the whole plant, this percentage includes those that are not occluded or that suffer from imperfections.

At the end of the segmentation you could see the edges of the colors that indicate the result of the "watershed algorithm" (Fig. 13).

Fig. 13. Targeting labels

The marked labels have been checked by using the trained network in order to it to determine according to the given characteristics, each segment that represents a leave for its later revision, finishing the algorithm (Fig. 14).

Fig. 14. Leaves classified by the network

When using the Watershed method certain difficulties were found, for example, if the leaves are superimposed and their textures are very similar the method assumes that there is only one leave present in that position (Fig. 15).

Fig. 15. Segmented and classified leaves

Another drawback that could cause problems when the algorithm gave its result, is the fact that the crop where the images were taken, was not organized by shrubs, causing a greater number of overlap between the leaves. Generally, these crops have separated bushes making easier to take cleaner and clearer photos to determine the different angles of the capture giving an optimal result and having a higher percentage of success.

4 Conclusions

An adequate background extraction was achieved allowing the image to work more easily and obtaining a cleaner close-up on which a better result could be searched at the time of the marker's location.

The morphological characteristics of the leaves helped in a great way to discard those results that are not of interest.

The neural network was designed with success, since it was able to correctly classify almost all those desired segments, recognizing the patterns that corresponded to the characteristics of a leaf. With this network a success of 92% was obtained. The estimation of this value does not tend to be so variable since the cross validation method was used in order to predict the fit of a model to a hypothetical set of test data.

Through the Watershed method it was possible to achieve an 8% success, allowing to separate the leaf from the plant in an almost individual way. It is a viable result to identify the state in which it finds itself at the moment the sample is taken, successfully identifying if the plant has at that moment pest samples.

In the leaves that contained holes, an adequate extraction and counting of these holes was achieved, this data is important in order to obtain an estimate of the state of the plant and the progress of the pest.

For future works, it will be designing an algorithm that allows improving the segmentation automatically, improving the method used. It is expected to test the algorithm in images taken in an organized crop varying scenarios, such as the moment of the growth phase of the crop being examined, as well as the angle of capture of the photo to determine in which of its images the detection is optimal.

References

1. Wang, Z., Wang, K., Yang, F., Pan, S., Han, Y.: Image segmentation of overlapping leaves based on Chan-Vese model and Sobel operator. Inf. Process. Agric. **5**(1), 1–10 (2018). https://doi.org/10.1016/j.inpa.2017.09.005
2. Gómez-Ossa, L.F., Botero-Fernández, V.: Application of artificial neural networks in modeling deforestation associated with new road infrastructure projects. Dyna **84**(201), 68–73 (2017). https://www.redalyc.org/articulo.oa?id=49650911008

Assessment of Different Approaches to Model the Thermal Behavior of a Passive Building via System Identification Process

Miguel Chen Austin[1(✉)], Ignacio Chang[1], Denis Bruneau[2], and Alain Sempey[3]

[1] Universidad Tecnologica de Panama, Avenida Domingo Diaz,
Panama City, Panama
{miguel.chen,ignacio.chang}@utp.ac.pa

[2] GRECCAU, EA, 748, ENSAP, Bordeaux, 33405 Talence, France
denis.bruneau@bordeaux.archi.fr

[3] Institute of Mechanical Engineering (I2M), UMR 5295, CNRS,
and University of Bordeaux, Arts et Metiers ParisTech, Bordeaux, France
alain.sempey@u-bordeaux.fr

Abstract. A preliminary study is presented with the aim of modeling the thermal behavior of a passive building that is ventilated merely with the promotion of natural ventilation, various models have been assessed by using the system identification process. The identification of a simplify and lite model of such thermal behavior is needed to later control the thermal comfort of the indoor environment through the building natural ventilation openings and window blinds. A physical-phenomena-based model using electrical analogies is built upon hypotheses allowed by the architectural features of the building. This helps analyze the interaction between the main elements of the physical domain, where the thermal behavior is only determined by the indoor air and concrete-slab temperatures. Three model approaches are examined with the help of the system identification toolbox: State space, Process models (linear and frequency domain), and Nonlinear representation. The nonlinear representation model is the best fitted encountered after 13 iterations with an accuracy of 71%.

1 Introduction

Due to the notary increasing energy consumption related to cooling in buildings, different low-consumption techniques are being developed or re-studied. The present work is focused on the enhancement of comfort management through passive or semi-passive solutions, such as free cooling. Such free cooling consists of the implementation of natural ventilation (part of such passive solutions), in this work, firstly, all interactions between the thermal mass of the building and the outdoor environment of a passive building are modeled, to understand its

A. Martínez et al. (Eds.): LACAR 2019, LNNS 112, pp. 185–193, 2020.
https://doi.org/10.1007/978-3-030-40309-6_18

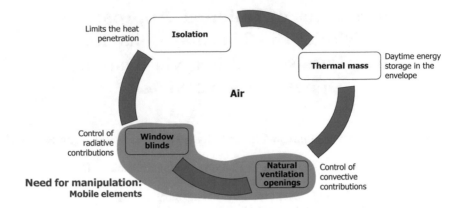

Fig. 1. Elements of the envelope in a passive building.

thermal behavior. Then, a simplified model is wanted to be identified by using the system identification procedure described in [2].

In general, the envelope of a passive building (such as nearly-zero-energy or plus-energy buildings) consists of four principal elements, which are (in decrease importance): Insulation, thermal mass, and the mobile elements (window blinds and openings) (Fig. 1). The insulation element helps to limit heat penetration through the building envelope. The thermal mass helps to store the incoming heat, attenuating its effects on the indoor environment. Finally, the mobile elements which refer to the window blinds and natural ventilation openings, help in principle to control the radiative and convective heat contributions to the indoor environment.

Since a passive building with such a definition can only rely on its mobile elements to manage the indoor thermal comfort apart from the thermal mass and insulation, the manipulation of these mobile elements is crucial. This manipulation must be represented by a real-time control process, and such type of control process needs the system to be represented by a simplified and light model. The most common approaches implemented to model such systems are the mere energy balances consider the nonlinearities introduced by the thermal buoyancy effects [7,9], and the electrical analogies through state-space representations since large systems of simultaneous differential equations can be manipulated easily [5,6,8].

2   Modeling the Thermal Behavior of the Building

The passive building employed in this study is a PEH prototype in Southwest France, named Sumbiosi. This prototype was designed in such a way as to promote the passive energy storage in winter daytime and the semi-passive energy discharge in summer nighttime.

The building consists of three main architectural features that aid to promote the strategies of passive energy charge and semi-passive energy discharge,

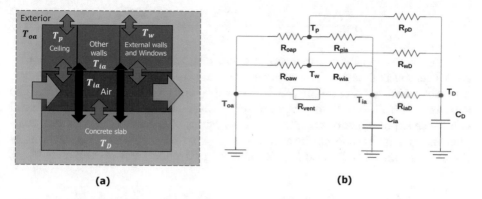

Fig. 2. (a) Domain modeling, and (b) Thermal model using electrical analogies.

respectively: a concrete-slab behind the South-facing facade with large double-glazed windows, fixed and programmable window blinds, and programmable natural ventilation openings at the South and North facades, and at the shed-roof. The East and West facades consist of an outer structure in maritime pine and an internal structure in wooden furnishes, with indoor and outdoor cabinets that serve as thermal buffers. The internal structure of the North, East and West facades, as well as the ceiling and floor, include a thermal insulation layer of 0.32 m thick. The South and North facades include the natural ventilation openings, where the area of the openings is 9.58% and 7.51%, respectively of the facade surface. The platform envelope encloses an air volume of about 211 m³, with an envelope surface of about 226 m².

The architectural features of this passive building allow considering some simplifications to the real physical domain (Fig. 2(a)), as follows:

- The heat is stored in the indoor air and the concrete slab, under the lumped-capacitance approximation, since the most heavy-weighted element of the thermal mass of the envelope is the concrete slab.
- Only the windows, main doors, and North and South facades are considered as external walls. Other walls, such as the indoor cabinets, partitions, and also, furniture and other equipment, are considered to be in thermal equilibrium with the indoor air.

These considerations allow us to suppose that the thermal behavior of the building can be determined only by the knowledge of two temperatures, and to represent the modeling domain in Fig. 2(a), through the electrical analogy of RC systems as in Fig. 2(b). From this RC thermal model presented in Fig. 2 (b), the following equations can be written:

$$C_{ia}\frac{dT_{ia}}{dt} = \frac{T_w - T_{ia}}{R_{wia}} + \frac{T_D - T_{ia}}{R_{Dia}} + \frac{T_p - T_{ia}}{R_{pia}} + \frac{T_{oa} - T_{ia}}{R_{vent}} \tag{1}$$

$$C_D \frac{\mathrm{d}T_D}{\mathrm{d}t} = \frac{T_{ia} - T_D}{R_{Dia}} + \frac{T_p - T_D}{R_{pD}} + \frac{T_w - T_D}{R_{wD}} \tag{2}$$

where C represents the thermal capacitance, T the temperature and R the equivalent thermal resistance. The subscript "ia" represents the indoor air, "D" the concrete slab, "p" the ceiling, "w" the walls, and "oa" the outdoor air. The subscript "$vent$" stands for "ventilation". These two principal equations need two other expressions to complement the system of equations. These two complementary expressions are introduced here as to relate T_p and T_w in terms of the output variables T_{ia} and T_D. By applying Kirchhoff law on the node T_p, it yields:

$$T_p = \left(\frac{R_{pD} R_{pia} R_{oap}}{R_{pia} R_{oap} + R_{pD} R_{oap} + R_{pD} R_{pia}} \right) \left(\frac{T_D}{R_{pD}} + \frac{T_{ia}}{R_{pia}} + \frac{T_{oa}}{R_{oap}} \right) \tag{3}$$

and for the node T_w:

$$T_w = \left(\frac{R_{wia} R_{wD} R_{oaw}}{R_{wD} R_{oaw} + R_{wia} R_{oaw} + R_{wia} R_{wD}} \right) \left(\frac{T_{ia}}{R_{wia}} + \frac{T_D}{R_{wD}} + \frac{T_{oa}}{R_{oaw}} \right) \tag{4}$$

by replacing expressions (3) and (4) into Eqs. (1) and (2), the system of equations is reduced to two differential equations with two unknowns: T_{ia} and T_D.

3   System Identification Modeling Procedure

The identification procedure employed here is the one proposed by [3], which consists of the following primary components:

- **Prior knowledge and objectives.** Presented and discussed here before in Sects. 1 and 2.
- **Experimental design for data collection.** A measurement campaign was conducted during the summertime 2016 in Southwestern France. During this campaign, several parameters were monitored as part of the experimental study on the coupling between natural ventilation and the building energy charge and discharge. Here, two parameters are of interest: the outdoor and indoor air temperatures. Both temperatures were collected at a sample rate of one observation per minute, for 34 days (Fig. 3).
  The experimental plan implemented during this campaign consisted of an unoccupied building; all window blinds were kept close as to limit the most the solar radiation gains, and the natural ventilation openings were programmed to open when the outdoor temperature falls below the indoor temperature (situation encountered mostly at night).
- **Model set.** Since the model set refers to the group of candidate models entirely determined by prior knowledge of the system behavior and characteristics, only three types of model are considered here: A 2nd order state space model, a two poles process model (linear model in frequency domain), and a nonlinear representation.

- **Criterion function and computation.** The method for the iterative minimization between the model result and the experimental data is based on the *fmincon* along with the sequential quadratic programming, with the help of the system identification toolbox in Matlab.

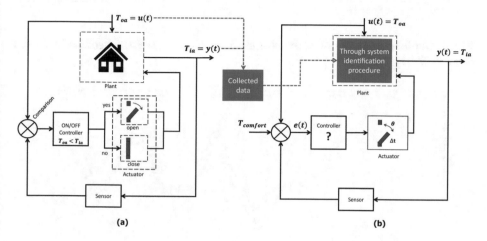

Fig. 3. (a) Experimental design and (b) System identification needs and objectives.

4 Results Analysis and Discussion

This section presents the simulation results after using the system identification toolbox and its comparison to the experimental data collected during the measurement campaigns.

Figure 4 shows the collected data for the outdoor air temperature. This temperature is considered as the only input variable to the system; named $u(t)$. The variability presented during the daytime (crests) is due to the perturbations introduced by the wind speed to the thermocouple measurement.

The indoor air temperature measurement is presented in Fig. 5 (black line). This temperature is considered as the only output variable of the system; named $y(t)$. Also, in Fig. 5 are presented the numerical results for the three models identified: Process model (dark red line), State space representation (yellow line), and the nonlinear representation model (green line).

4.1 Comparison of the Identified Models

In Fig. 5, it can be observed that the best-fitted model is the nonlinear representation (with an accuracy of 71.61%), followed by the process control model (with an accuracy of 62.81%), and the state space representation model (with an accuracy of 61.14%).

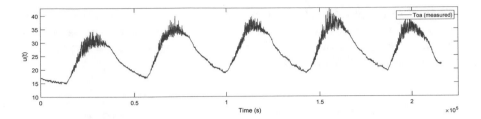

Fig. 4. Outdoor temperature measurement collected during the summertime 2016 from August 12th to 17th.

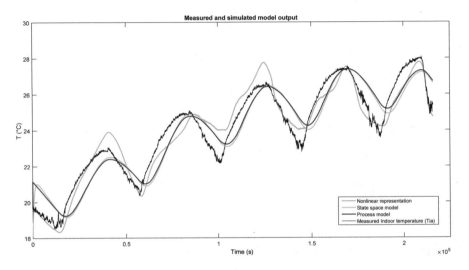

Fig. 5. System identification results with decreasing accuracy of: 71.61% (Nonlinear representation), 62.81% (Process model), and 61.14% (State space).

From the point of view of the presented numerical results, the dynamics described by the nonlinear representation model adapt best to the measured indoor temperature in terms of the time-shifting and starting behavior. The time-shifting (known as the thermal lag in building science) refers to the time length difference between two temporal curves. It can be observed that the green curve shows more coincidences with the valleys of the black curve, while the red and yellow curves always show a temporal shift of about 0.03×10^5 s (almost one hour) between the experimental and models values. This time shift might be explained by the starting behavior of these last two numerical models iterations. Here, the starting behavior refers to the trend in the slope, adopted by the numerical model, at the beginning of the iterations. In Fig. 5, only the slope of the nonlinear model (green line) appears to follow the negative slope (or decreasing behavior) presented by the experimental value (black line). The other two (red and yellow lines) models, present a positive slope (or increasing behavior) at the beginning.

From the point of view of the physical phenomena involved and the type of models examined, the state space and process models describe best the behavior of linear systems in the frequency domain. However, the knowledge of the governing equations of the physical phenomena involved in the process of naturally ventilating buildings indicates that a nonlinear behavior should be expected. Such physical phenomena involved are the thermal buoyancy effects [4], wind effects, and thermal radiation. A simplify model showed that a satisfactory enough accuracy could be accomplished when accounting explicitly for all these nonlinearities [1].

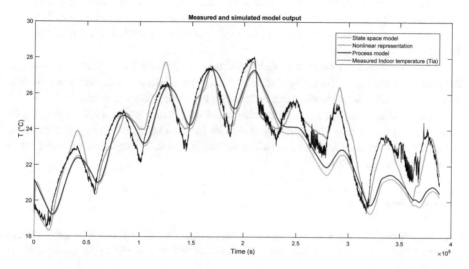

Fig. 6. System identification results with decreasing accuracy of: 62.79% (Nonlinear representation), 38.81% (Process model), and 31.75% (State space).

4.2   Candidate Model and Validation Data

As it is encountered, the thermal behavior represented by the indoor air temperature of the building is best modeled by a nonlinear representation. This finding, as explained before, is based on the notion of the physical phenomena involved in the process of naturally ventilating a building.

In order to validate the nonlinear representation model, another set of collected data (the following days of the data presented in Fig. 5) is chosen and the simulation is ran again for the entire data set (Fig. 6). For this validation, all three models remained the same as encountered for the working data presented in Figs. 4 and 5.

In Fig. 6, it can be observed that the nonlinear representation best models the measured indoor air temperature with an accuracy of 62.79%.

In this preliminary study, for these three types of models examined, a model that includes the nonlinearities in this system is the best representation of the thermal behavior of a passive building and should be considered as the candidate model.

5   Conclusion and Future Work

In this preliminary study, the thermal behavior of a passive building is wanted to be modeled, considering that its behavior is represented only by the indoor air temperature. This building is ventilated naturally. Three classic models were examined to model the thermal behavior of such building: Process model, State-space model, and nonlinear representation, determined with the help of the system identification toolbox.

Experimental data collected during the summertime 2016 was first used to determine these three models, and another part of these data set was used to validate the candidate model: the nonlinear representation model.

These findings encourage the development of further work, which will be dedicated to implementing this nonlinear model in the design of a proper controller to manage the indoor thermal comfort through the manipulation of the mobile elements of the envelope.

References

1. Chen Austin, M.: On the coupling between natural ventilation and sensible energy charge and discharge in buildings: an experimental and modeling approach. Ph.d. thesis
2. Keesman, K.J.: System Identification: An Introduction. Advanced Textbooks in Control and Signal Processing. Springer, London (2011). OCLC: ocn747112444
3. Ljung, L.: Analysis of a general recursive prediction error identification algorithm. Automatica **17**(1), 89–99 (1981). https://doi.org/10.1016/0005-1098(81)90086-8. https://linkinghub.elsevier.com/retrieve/pii/0005109881900868
4. Qingyan Chen, Y.J.: Buoyancy-driven single-sided nautral ventilation in building with large openings. Int. J. Heat Mass Transf. **46**, 973–988 (2003)
5. Tang, R., Wang, S.: Model predictive control for thermal energy storage and thermal comfort optimization of building demand response in smart grids. Appl. Energy **242**, 873–882 (2019). https://doi.org/10.1016/j.apenergy.2019.03.038. https://linkinghub.elsevier.com/retrieve/pii/S0306261919304441
6. Viot, H., Sempey, A., Mora, L., Batsale, J.C.: Fast on-site measurement campaigns and simple building models identification for heating control. Energy Procedia **78**, 812–817 (2015). https://doi.org/10.1016/j.egypro.2015.11.107. http://www.science direct.com/science/article/pii/S1876610215018391
7. Yang, D., Guo, Y.: Fluctuation of natural ventilation induced by nonlinear coupling between buoyancy and thermal mass. Int. J. Heat Mass Transf. **96**, 218–230 (2016). https://doi.org/10.1016/j.ijheatmasstransfer.2016.01.017. http://www.scie ncedirect.com/science/article/pii/S0017931015312369

8. Yang, S., Wan, M.P., Ng, B.F., Zhang, T., Babu, S., Zhang, Z., Chen, W., Dubey, S.: A state-space thermal model incorporating humidity and thermal comfort for model predictive control in buildings. Energy and Buildings **170**, 25–39 (2018). https://doi.org/10.1016/j.enbuild.2018.03.082. https://linkinghub.elsevier.com/retrieve/pii/S0378778817340215
9. Zhou, J., Zhang, G., Lin, Y., Li, Y.: Coupling of thermal mass and natural ventilation in buildings. Energy Build. **40**(6), 979–986 (2008). https://doi.org/10.1016/j.enbuild.2007.08.001. http://linkinghub.elsevier.com/retrieve/pii/S0378778807002083

A Fast Solution to the Dual Arm Robotic Sequencing Problem

Francisco Suárez-Ruiz[1] and Carol Martinez[2]($\boxtimes$)

[1] nuTonomy Asia, 77 Ayer Rajah Crescent, 01-30, Singapore 139954, Singapore
[2] Department of Industrial Engineering, Pontificia Universidad Javeriana,
Bogotá, Colombia
carolmartinez@javeriana.edu.co

Abstract. Robotics applications such as spot-welding, spray-painting, drilling, and objects handling, require the robot to visit successively multiple targets. The robot travel time among targets is a significant component of the overall execution time. This travel time is in turn greatly affected by the visiting order and by the robot configurations used to reach each target. It is crucial to optimize these elements during the motion planning stage, a problem known in the literature as the Robotic Task Sequencing Problem (RTSP).

This problem has been studied by the robotics community but only for single arm manipulators. Our contribution in this paper is to extend an existing RTSP algorithm to the dual arm case. The key to our approach is to exploit the classical distinction between task-space and configuration-space, which, surprisingly, has been so far overlooked in the RTSP literature.

We show experimentally that our method finds high-quality motion sequences and requires a reduced amount of computation time, making it ideal for real-world applications. For a toy-example of a pick-and-place task, the proposed dual-arm *RoboTSP* takes less than 20 s to solve the task sequencing problem and compute the collision-free trajectories.

1 Introduction

Robots work together with factory employees to manufacture goods with high quality standards while lowering the production costs. These robots already make our lives better by tackling tasks that are dangerous, repetitive, tedious or boring, and give us added accuracy, precision, and strength. Moreover, robots increase our productivity and help us to accomplish more in a world were the working population is getting older.

Despite that, robotics applications such as spot-welding, spray-painting, drilling, and objects handling where the robot is required to visit multiple targets, are constrained to carefully engineered environments. There, the robots can be programmed once, using the classical teach-and-replay strategy commonly used in traditional robotic automation to follow repetitive and well defined motion sequences.

A. Martínez et al. (Eds.): LACAR 2019, LNNS 112, pp. 194–202, 2020.
https://doi.org/10.1007/978-3-030-40309-6_19

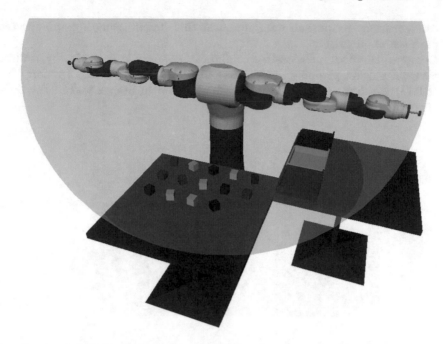

Fig. 1. Pick-and-place classification task. The objective is to pick-and-place all the cubes into their corresponding color bins. The reachable workspace of the bimanual robot has a ellipsoid shape.

Consider for instance the pick-and-place task depicted in Fig. 1 where the robot is required to classify objects of different categories, each category corresponding to an unique color. The robot travel time between cubes and bins is the most significant component of the overall execution time. This travel time is in turn greatly affected by the visit order of the cubes and by the robot configurations used to reach the cubes, either using single or dual arm configurations. The optimization of these elements is a problem known in the literature as the RTSP, see [1] for a recent review.

In [8], the authors present a fast, near-optimal, algorithm to solve the Robotic Task Sequencing Problem for single arm robots. The key to their approach is to exploit the classical distinction between task-space and configuration-space, which, surprisingly, has been so far overlooked in the RTSP literature.

Specifically, they propose a three-step algorithm [see Fig. 2 for illustration]:

1. Find a (near-)optimal visit order of the targets in a *task-space metric* (e.g. Euclidean distance between the targets), using classical Traveling Salesman Problem (TSP) algorithms.
2. Given the order found in Step 1, find for each target the optimal robot configuration, so that the total path length through the configurations is minimized in a *configuration-space metric* (e.g. Euclidean distance between the robot

configurations – collisions are ignored at this stage), using a graph shortest path search algorithm.
3. Compute the final collision-free configuration-space trajectories by running classical motion planning algorithms (e.g. Rapidly-Exploring Random Tree (RRT)) through the robot configurations found in Step 2 and in the order given by Step 1.

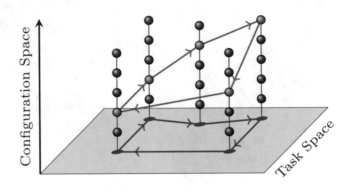

Fig. 2. Decomposition of the Robotic Task Sequencing Problem into task-space and configuration-space. The blue circles in the task-space correspond to the target objects. The visit order is found in Step 1 of the algorithm. The black spheres represent the IK solutions (robot configurations) at which the robot is able to reach the corresponding target object. Finally, in Step 2, the algorithm selects the best robot configurations (green spheres) to minimize the task execution time.

Moreover, they carefully benchmark different key components of the *RoboTSP* algorithm, such as underlying task-space TSP solver, configuration-space metrics and discretization step-size for the free-DoFs, so as to come up with an efficient solution.

Our contribution in this work is to extend the *RoboTSP* algorithm to support dual arm manipulators. In order to achieve this, we also exploit the distinction between task-space and configuration-space. First, we find all the dual arm configurations that would allow the robot to pick all the cubes, in pairs, and place them in their corresponding bins. This is done prior to Step 1 of the *RoboTSP* algorithm and from there we continue with the original steps as described in Sect. 2.

The remainder of the paper is organized as follows. In Sect. 2, we introduce and describe in detail the considerations and modifications that should be applied to the *RoboTSP* in order to support dual arm RTSP. In Sect. 3, we present the experimental results, showing that the proposed method finds efficient motion sequences with computation times that are practical for real-world applications. Finally, in Sect. 4, we conclude with few remarks.

2 RoboTSP for Dual Arm Manipulation

For the single arm case, the distinction between task-space and configuration-space is straight forward, given that there is a 1-to-1 matching between the task-space pose of the target object and the robot configuration that can reach it. For the dual arm scenario, the matching can also be 2-to-1, since one robot configuration can reach two target objects at once.

For simplicity, consider the situation where the robot is able to reach all the cubes, two at a time, that is ts targets in the task-space that can be reached by $n = \frac{ts}{2}$ robot configurations. A tour in the task-space that visits each target exactly once is called a *task-space tour* [1]. We first compute IK solutions for each target pair – using a suitable discretization for the free-DoFs. In the task at hand, shown in Fig. 1, we have two free-DoFs per arm because the pick-and-place actions only have 5-DoF and each arm has 7-DoF.

A tour in the configuration-space that starts from the robot "home" configuration, visits, for each target pair, exactly one IK solution associated with that target pair, and returns to the "home" configuration is called a *configuration-space tour*. Our objective is to find the fastest *collision-free* configuration-space tour subject to the robot constraints (e.g. velocity and acceleration bounds).

Let m_i be the number of IK solutions found for a target pair i. If we do not take into account obstacles, there are $n!(\prod_{i=1}^{n} m_i)$ possible configuration-space tours (with straight paths) for this task. One cannot therefore expect to find the optimal sequence by brute force in practical times.

2.1 Proposed Algorithm

As discussed in Sect. 1, here we propose a method to extend the *RoboTSP* algorithm for dual arm manipulation. The steps of the modified algorithm are:

1. Finding a feasible task-space tour. Here we compute all the IK solutions for each target pair and find a attainable combination of robot configurations such that every target is visited only once by any end-effector. This yields an ordered sequence of target pairs.
2. Given the target pairs order found in Step 1, finding the optimal robot configuration for each target pair that minimizes the corresponding configuration-space tour length in a *configuration-space metric*. In order to reduce computational time, collisions are ignored at this stage.
3. Computing fast collision-free *configuration-space* trajectories by running classical motion planning algorithms (e.g. RRT-Connect with post processing [4,7]) traversing the robot configurations found in Step 2 and in the order given by Step 1.

Steps 1 and 3 can be solved using off-the-shelf tools. Furthermore, their implementation details and benchmarking results are given in [8].

---

[1] Strictly speaking, a tour requires to return to the first target, so we are making a slight abuse of vocabulary here.

Regarding Step 2, we first construct an undirected graph as depicted in Fig. 3. Specifically, the graph has n layers, each layer i contains m_i vertices representing the m_i IK solutions of target pair i (the target pairs are ordered according to Step 1), resulting in a total of $\sum_{i=1}^{n} m_i$ vertices. Next, for $i \in [1, \ldots, n-1]$, we add an edge between each vertex of layer i and each vertex of layer $i+1$, resulting in a total of $\sum_{i=1}^{n-1} m_i m_{i+1}$ edges. Finally, we add two special vertices: "Start" and "Goal", which are associated with the robot "home" configuration, and connected respectively to the m_1 vertices of the first layer and the m_n vertices of the last layer.

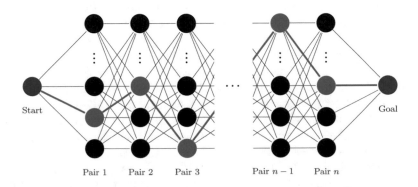

Fig. 3. Graph constructed for Step 2 of the algorithm. The target pairs are ordered in Step 1. The shortest path (green lines) connecting the "Start" and "Goal" vertices will yield the optimal configuration-space tour that visits the pairs in the order specified by Step 1.

The *configuration-space metric* that defines the edge cost in Step 2 of our algorithm is a key component for the overall performance of the method. Given two robot configurations $\boldsymbol{q}$ and $\boldsymbol{q}'$, the ideal cost of the edge $c^*(\boldsymbol{q}, \boldsymbol{q}')$ is the duration of a time-optimal collision-free trajectory between them. However, since the costs must be computed for all $m_1 + \sum_{i=1}^{n} m_i m_{i+1} + m_n$ graph edges, running full-fledged motion planning algorithms (with collision checks) for every edge would not be tractable. Therefore, one must consider approximate metrics that are fast to compute, yet give a good prediction of the corresponding time-optimal collision-free trajectory duration. Here we use the metric suggested in [8], the *Maximum joint difference* where the edge cost $c(\boldsymbol{q}, \boldsymbol{q}')$ is estimated as follows:

$$c(\boldsymbol{q}, \boldsymbol{q}') := \max_k \left| \frac{(q_k' - q_k)}{\dot{q}_k^{\max}} \right|,$$

where $\dot{q}_k^{\max}$ is the maximum joint velocity for joint k. The intuition of this metric is to determine the maximum joint displacement when "moving" from $\boldsymbol{q}$ to $\boldsymbol{q}'$ by simply computing the joint difference, $(q_k' - q_k)$, over the maximum joint velocity, for $k \in [1, \ldots, \text{DoF}]$.

Finally, we run a graph search algorithm to find the shortest path between the "Start" and "Goal" vertices. By construction, any path between the "Start" and "Goal" vertices will visit exactly one vertex in each layer, in the order specified by Step 1. Conversely, for any choice of IK solutions for the n target pairs, there will be a path in the graph between the "Start" and "Goal" vertices and going through the vertices representing these IK solutions. Therefore, Step 2 will find the *true optimal* selection of IK solutions that minimize the total cost, according to the specified configuration-space metric.

2.2 Complexity Analysis

For Step 1, it is well-known that TSP is NP-complete. It means that finding the true optimal tour for n target pairs has in practice an exponential complexity.

For Step 2, let M be an upper-bound of the number of IK solutions m_i per target. The number of graph vertices is then $\mathcal{O}(nM)$ and the number of the graph edges is $\mathcal{O}(nM^2)$. Since Dijkstra's algorithm [3] (with binary heap) has a complexity in $\mathcal{O}(|E| \log |V|)$ where $|E|$ and $|V|$ are respectively the number of edges and vertices, Step 2 has a complexity in $\mathcal{O}(nM^2 \log (nM))$.

For Step 3, one has to make $n - 1$ queries to the motion planner, yielding a complexity in $\mathcal{O}(n)$. However, as the constant in the $\mathcal{O}()$ (average computation time per motion planning query) is large, the overall computation time is dominated by that of Step 3 in our setting. In general, the computation time of motion planning queries depends largely on the environment (obstacles). See [6,8,9] for recent benchmarking results showing the CPU time required when planning practical robot motions.

3 Experiments

This section evaluates the proposed task sequencing method when applied to the pick-and-place classification task shown in Fig. 1. The dual arm robot used is a Yaskawa SDA10F 15-DoF industrial robot equipped with two suction cups. This is a modified version of the robot described in [5].

3.1 Whole-Body IK Solver for the SDA10F Robot

Each SDA10F robot arm has 7-DoF. Let us consider only the contact point between a suction cup and the top face of a cube. This contact point can be fully defined in 5D, that is, the cube XYZ position and the normal of the top face. This results in two free-DoFs, also known as redundant DoFs, for each arm. Moreover, the robot is equipped with and extra DoF at the torso joint, that rotates the torso and the two arms, all together.

The task at hand has $ts = 14$ cubes, that is, $n = \frac{ts!}{(ts-2)!} = 182$ possible cube pair permutations. In order to compute whole-body IK solutions within practical time bounds, several simplifications have been incorporated to develop the 15-DoF IK solver. First, we perform a *lazy* check to discard *crossed-arms*

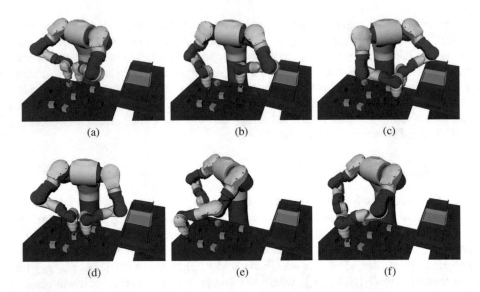

Fig. 4. Feasible IK solutions for picking up six different pairs of cubes. These solutions also are feasible for dropping the cubes in their corresponding color bin.

robot configurations because the arms collide with each other. Next, we compute the torso joint value that points towards the middle 3D point of the cube pair. Finally, we use OpenRAVE's IKFast [2] to solve the 5D IK problem for each arm. Figure 4 shows one feasible IK solution for picking up six different pairs of cubes.

For this task configuration, our solver finds an average of 305.4 feasible whole-body IK solutions for each cube pair, that is 2138 nodes for the configuration-space graph constructed in Step 2 of the dual arm *RoboTSP* algorithm.

3.2 Planning and Task Execution Times

The times presented in this section were measured in a system with Intel® Core™ i7-8750H CPU @ 2.20 GHz x 12, 16 GB RAM, GeForce GTX 1060 video card, and running Ubuntu 16.04 (Xenial), 64 bits.

Table 1. Planning time for each step of the dual arm *RoboTSP* algorithm when applied to the task at hand.

Step	Summary	Duration [s]
1	IK solutions and cube pairs visit sequence	2.16
2	Optimal robot configuration for each target pair	3.21
3	Collision-free trajectories	3.79
Total planning time		9.16

Table 1 depicts the planning time required for each step of the method proposed in this work. We can note that the computation times reported in previous works, see [8] for more details, are several orders of magnitude higher than ours.

Finally, the task execution time achieved using the dual arm *RoboTSP* method for the task at hand is 18.41 s. A visualization with the trajectories produced to pick and place the 14 cubes is available at https://youtu.be/3zoiKgTFv0s.

4   Conclusions

In this paper, we have propose a method to find a near-optimal sequence to visit n target pairs, using a dual arm robot, with multiple whole-body configurations per pair, a problem known in the literature as the Robotic Task Sequencing Problem. For a toy-example of a dual arm pick-and-place task which requires visiting 7 target pairs, that is, 14 targets, our method computes a high-quality solution in less than 20 s. To our knowledge, no existing approach could have solved this problem within practical times for real-world applications.

Future works will include the implementation of the dual arm *RoboTSP* algorithm in the real robot described in [5] and its application towards 6D object classification tasks. The robot end-effectors are, a parallel-jaw gripper and, a 3-finger adaptive gripper. We foresee that the robot redundancy with respect to the pick-and-place task needs to be exploited furthermore in order to maintain the computation times of our method under practical bounds.

Acknowledgement. This work was supported in part by Pontificia Universidad Javeriana, Bogota, Colombia, under the grant "Becas de Movilidad - 2018".

References

1. Alatartsev, S., Stellmacher, S., Ortmeier, F.: Robotic task sequencing problem: a survey. J. Intell. Robot. Syst. **80**(2), 279–298 (2015). https://doi.org/10.1007/s10846-015-0190-6
2. Diankov, R.: Automated construction of robotic manipulation programs. Ph.D. thesis, Carnegie Mellon University, Robotics Institute (2010)
3. Dijkstra, E.W.: A note on two problems in connexion with graphs. Numer. Math. **1**(1), 269–271 (1959). https://doi.org/10.1007/BF01386390
4. Kuffner, J., LaValle, S.: RRT-connect: an efficient approach to single-query path planning. In: IEEE International Conference on Robotics and Automation (ICRA), vol. 2, pp. 995–1001 (2000). https://doi.org/10.1109/ROBOT.2000.844730
5. Martinez, C., Barrero, N., Hernandez, W., Montaño, C., Mondragón, I.: Setup of the Yaskawa SDA10F robot for industrial applications, using ROS-industrial. In: Chang, I., Baca, J., Moreno, H.A., Carrera, I.G., Cardona, M.N. (eds.) Advances in Automation and Robotics Research in Latin America, pp. 186–203. Springer (2017)
6. Meijer, J., Lei, Q., Wisse, M.: Performance study of single-query motion planning for grasp execution using various manipulators. In: 2017 18th International Conference on Advanced Robotics (ICAR), pp. 450–457. IEEE (2017). https://doi.org/10.1109/ICAR.2017.8023648

7. Pham, Q.C.: Trajectory planning. In: Handbook of Manufacturing Engineering and Technology, pp. 1873–1887. Springer, London (2015). https://doi.org/10.1007/978-1-4471-4670-4_92
8. Suárez-Ruiz, F., Lembono, T.S., Pham, Q.: RoboTSP - a fast solution to the robotic task sequencing problem. In: 2018 IEEE International Conference on Robotics and Automation (ICRA), Brisbane, Australia (2018)
9. Sucan, I.A., Moll, M., Kavraki, L.E.: The open motion planning library. IEEE Robot. Autom. Mag. **19**(4), 72–82 (2012). https://doi.org/10.1109/MRA.2012.2205651

mHealth System for the Early Detection of Infectious Diseases Using Biomedical Signals

José Sanz-Moreno[1], José Gómez-Pulido[2(✉)], Alberto Garcés[1],
Huriviades Calderón-Gómez[2], Miguel Vargas-Lombardo[3],
José Luis Castillo-Sequera[2], María Luz Polo Luque[2], Rafael Toro[1],
and Gloria Sención-Martínez[4]

[1] University Hospital Principe de Asturias, Madrid, Spain
josesanzm@gmail.com, albertogarces0@gmail.com, rafael.toro@uah.es
[2] University of Alcalá, Madrid, Spain
{jose.gomez,jluis.castillo,mariluz.polo}@uah.es
[3] Universidad Tecnológica de Panamá, Panama City, Panama
miguel.vargas@utp.ac.pa
[4] Autonomous University of Santo Domingo, Santo Domingo, Dominican Republic
dragloriasencion@gmail.com

Abstract. Early detection of infectious diseases is a major clinical challenge. When diagnosis comes after symptoms has a bad effect in health, but also spread a contagious approach towards other people. The proposed e-Health system supports the pre-diagnosis of these diseases. It gathers vital signs simultaneously (Electrodermal Activity, Body Temperature, Blood Pressure, Heart Beat Rate and Oxygen Saturation) from residents with a portable and easy-to-use biomedical sensors kit and managed with an Android App once a day. The processed data is uploaded to an online database for being used as SaaS to build the predicting models. The mHealth system may be operated by the same personnel on site not requiring to be medical or computational skilled at all. A real implementation has been tested and results confirm that the sampling process can be done very fast and steadily The same experiment showed that the manipulation of the App had a fast learning curve and no significant differences are observable in learning time by people with different skills or age. These usability factors are key for the mHealth system success.

1 Introduction

Nowadays, societies are ageing causing the increment of the elderly and depending people amount [10]. This fact increases the medical assistance costs and pathology ordering. Automate monitoring and pre-diagnosis of diseases could improve the life standards of these patients and reduce health costs. This project is inspired by the fact that this approach was already working for high risk and

A. Martínez et al. (Eds.): LACAR 2019, LNNS 112, pp. 203–213, 2020.
https://doi.org/10.1007/978-3-030-40309-6_20

very severe cardiopathies or pneumopathies [6,13], however it was not practicable, nor sustainable for respiratory and urinary infections, due to the high prevalence of cases.

e-health does not only enhance the classical approach of medicine but brings new paradigms that require attention. The pharmacological industry benefits from more realistic tests that burst the efficacy of the research process [7]. Another example is the participating medicine or patient-centric, based in the communication, empathy and collaboration between patient and practitioner. Doctors get patients to actively participate in the treatment, getting more mutual satisfaction [1].

Smartphone technology evolution has helped to decrease transcription errors and has made that biomedical data are available everywhere. On the other hand, healthcare personnel becoming familiarized increasingly with mobile applications (Apps), introducing them for different purposes such as drug reference search, disease diagnosis, medical calculator, and pregnancy wheel [18].

Cloud is advantageous as hides the technicalities and brings scalability. However, it requires attention to implement systematic security model approach [8]. It is also necessary to indicate that the fact of adding more functionality to the cloud platform, however does not show that the acceptance is immediate in clinical activity [3]. An extra effort of communication around any technical implementation is necessary to minimize the disruption. The issue of personal data and privacy must be always addressed as it normally becomes a barrier [21].

The online DB can be also monitored in real time for Telecare, so that immediate action can be taken when any change raises an alert [20]. One of the key concerns is to design good filtering policies so that keeping at a minimum the raising of false alarms. Remote care, or Telecare brings more effectiveness providing real time surveillance and releasing medical specialists, becoming more focused on their main tasks with their patients.

Finally, the software architecture is also important to develop different and unforeseen services that consume the data to provide expert systems or monitoring alert systems based in accurate data models [5,11,15,19].

It is the first achievement of a wider international research initiative carried out by EU Member States and Associated Countries and the Community of Latin American and Caribbean States (CELAC) within the 7th Framework Programme for Research and Technology Development (FP7). The name of the project is the "Design and implementation of a low-cost smart system for pre-diagnosis and telecare of infectious diseases in elderly people", abbreviated SPIDEP.

This study presents an e-Health system that responds to the difficulty of getting relevant data for predicting infectious diseases. The system comprises a biomedical sensor kit, a mobile APP, a SaaS database, a microservice software architecture and a procedure for practitioners to take measurements.

It is important to observe that many similar systems are intended for home monitoring, minimizing health costs, but the proposed system focus on the field

of institutionalized residents care. This brings two important consequences that
have not been treated that much before. First one is about the difficulties for
practitioners to get used to the new systems, normally seen as an added workload
and second one is about the unnecessary skills required for residents to manage
the platform [14]. The similarities that the new procedure has with the existing
standard procedures and the easy usability make the deployment to be very fast
and keep the necessary warm relation of nurses with residents [17]. Additionally,
this fast deployment speed maximizes the quality of data, improving the accuracy
of the predicting models.

The biomedical sensor kit allows to gather the vital signs of residents in a
short fraction of time, as it is proved in the study. The kit is operated with
an Android APP, currently designed for a standard tablet, that collects and
stores vital signs to upload them to the Cloud database to visualise and manage
the medical and contextual information. The SaaS database is available online
continuously updated with the right credentials, proportional to the required
special protection of medical data. The software is implemented in a microservices architecture of several layers that provide the necessary functionality for
the future e-Health applications and for the data management. This architecture also allows the consumption of data either by users via user interfaces or
application via APIs, being each piece of functionality being unlimited scalable.

Section 2 describes the system components and their particular configuration
for the process. Section 3 explains the functionality of the model. Section 4 shows
the results and Sect. 5 shows the conclusions and foreseen research lines.

2   Design and Functionality of the mHealth System

This Section describes the components of the proposed e-Health system for gathering data to predict models of infectious diseases. Figure 1 shows a schema of
the existing devices of the considered system. It consists of a portable hub, sensors kit, an Android mobile device running the App, local DB, online DB, and
a dedicated web portal.

Fig. 1. A schema of the existing devices of the considered system.

2.1   Biomedical Sensor Kit

This is the hardware with the biomedical sensors that are deployed daily on residents' body. The biomedical sensors are connected to a hub that multiplexes the concurrent measures collected.

The equipment of BiosignalsPlux (BPX) [4] was selected for its versatility, ease of use, battery duration and parallel data acquisition. The goal was also to take measurements very fast [9]. The medical team decided the vital signs to be recorded for predicting infectious diseases. They were body temperature, electrodermal activity (EDA), pulse oximeter, heart peat rate and blood pressure. The electrocardiogram sensor could be included in a further version to also allow the pre-diagnosis of hearth diseases, although would require special considerations due to the continuous nature of the signal. The kit was then adapted for this particular allocating the tablet and hiding the cables to facilitate its manipulation. Figure 2 is a picture of the system.

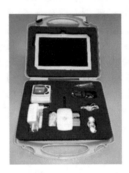

Fig. 2. Adapted case for carrying biosensors and tablet.

2.2   Data Manager APP

The second element is the mobile application, running on an Android-based tablet, which gathers and pre-processes the data from the hub deployed on the field. It gathers the vital signs from patients and pre-processes the signals into quantified values. It then checks that the taken measures are correct. The user then stores the values and uploads them when the tablet is connected to Internet. The APP can also work offline making the user not to stop the process. Additionally, the App establishes the connection with BPX device via Bluetooth protocol automatically.

2.3   Online DB

The third element is the Clinical Cloud DB accessible via web by authorized users that collects the information from all the APPs The online DB is the source of knowledge. It is filled up with the data sent by the APPs running on the field. There is no limitation in the number of APPs feeding up the DB. It is accessible via WEB with the user ID and password granted by the Administrator. SaaS authorized consumers have only-read privileges, so that they are not allowed for data change or deletion. The online DB collects the values of the biosensors, labelling with the patient id code, the institution id, the date, time and two flags, one to indicate whether the data was uploaded and the other should the record was deleted by the operator. The DB is designed to save both the data and the control parameters "Uploaded Flag" (UF) and "Deleted Flag" (DF). The UF parameter is set when a measure saved has been uploaded to the online DB. The DF parameter is set when the operator deletes the measure.

2.4 Microservices Software Architecture

There are two endpoints for devices (Mobile – API Gateway & PC – Web UI) to get access. Once entered users call the microservices with their credentials (nurse, doctor or administrator). Services are interconnected, presented as a single application for the whole process. Thus, microservices are grouped in three sets, sharing the same database. First microservices group deals with measurements from medical sensors. Second group manages residents, institutions, roles and privileges and doctors feedback data. Third group records infectious diseases and access to pre-diagnostic service. First and second groups replicate the database asynchronously and the third group of microservices, synchronously.

The architecture uses REST API for integrating, transferring and storing the data. Microservices software applications are split up in small services, interconnected throughout the Infrastructure Layer, becoming robust against failures and performing well in Cloud environments [12].

3 Application Design and Operation

This eHealth App has been designed to be simple, fast to learn, and easy-to-use by non-skilled personnel. The App blinds the measure values to patient and operator. Thus, no immediate intervention is not allowed, and the App does not stop collecting information. Additionally, according to privacy protection from international standards [16], patients are not recognized and only an "Id" is required. As shown in Fig. 3.

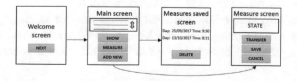

Fig. 3. Application use schema.

- *Welcome Screen: design and operation*
  Welcome Screen contains the logo and "NEXT" button to load the "Main Screen". On the first run, the App requests the institution name.
- *Main Screen: design and operation*
  Before loading the "Main Screen", the App checks the measures saved pending to be uploaded and reports it. Additionally, it contains a scrollable list of stored patients. When the patient is selected, buttons "MEASURE" and "SHOW" are enabled. New patients are added with "ADD NEW" button. On the other hand, patients registered cannot be deleted to avoid missing data.

Measures can be taken offline as data can be saved in the local DB. Pending measures saved can be upload to the online DB. The App communicates with the online DB via the required DB connector.

- *Measures Saved Screen: design and operation*
  Measures Saved Screen provides a list of date and time information of each measure, blinding the values. "DELETE" button deletes the selected measure from the list, but not from the local DB due safety reasons, setting the DF.
- *Measure Screen: design and operation*
  BPX device is connected automatically when it is switched on. Once the sensors are placed on the patient, the operator can start data transfer with "TRANSFER" button. Data transfer duration is set to 10 s by default, which means, 1 frame per second as a sample rate. The frames with the embedded data read by the sensors are received, processed and saved in temporary arrays. Temperature values stabilization can take a duration longer than 10 s since the App waits until a stable value is reached. Hence, the data transfer time established by default can be extended. Temperature stable value is computed by comparing last received value with the previous value saved, stopping the process once the tolerance is reached. The lower tolerance is configured; the more time is needed to get stable values based on (1).

$$|current\ value - old\ value| < tolerance \tag{1}$$

Figure 4 shows an example of temperature value and processing time necessary up to a stable value is found. In this example, the temperature reaches a stable value in 35 s.

Each data saved are then checked whether it fit the human ranges for every magnitude. The decision ranges are showed in Table 1.

When more than 50% of the received frames are out of range, an error icon appears and the other measures cannot be saved. When all sensors provide valid magnitudes the "SAVE" button is enabled and all measures can be saved. As shown in Fig. 5.

When the operator presses on the "SAVE" button, the App re-processes the array of frames for interesting values from all active sensors and saves them into local DB, adding the patient ID, date, and time information. The data obtained for each sensor is shown in Table 2.

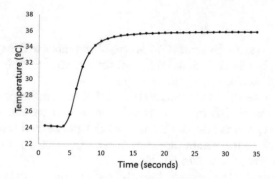

Fig. 4. Temperature stable value reaching.

Table 1. Magnitude ranges.

Sensor	Valid measure	Wrong measure
Temperature	34 °C–42 °C	<34 °C and >42 °C
EDA	≥0.2 µS	<0.2 µS
Oximeter	70%–100%	<70%
Pulse	≥30 bpm	<30 bpm
SYS pressure	≥30	<30
DIA pressure	≥60	<60

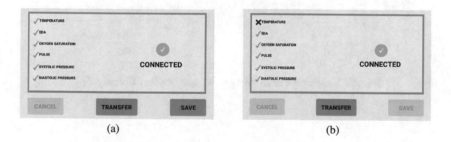

Fig. 5. Examples of correct (a) and wrong (b) measures.

Table 2. Re-process obtained values.

Sensor	Data obtained
Temperature	Stable temperature value
EDA	Min, Max, Average
Oximeter	Min, Max
Pulse	Min, Max, Average
SYS pressure	Last value saved
DIA pressure	Last value saved

4   Results

Since Android Operating System (OS) is one of the most widely used with 86.1% of market share [2], the App was developed for running on this OS. This makes the system very versatile.

The medical personnel have checked the BPX sensors with certified medical equipment. The online DB info is correctly visualized on the Web portal. The usability of the App was tested through a trial performed with 18 volunteers practicing once or twice whom age ranges was from 20 to 70 years and their mobile familiarity was from 2 to 5 scale.

The engineer explained how to place the sensors on the patient's bodies, how to use the App to collect the data, and, finally, how to upload the measures to the DB. Training task took on average 7 min. After the training, the deployment of the sensors on the patient took 1 min and 55 s. Total time required for taking an average measure of 4:15 min with a dispersion of ±2:21 min.

In Fig. 6a we can observe that the learning process is nearly uncorrelated with the age of the volunteers, giving R2 = 0.05. Figure 6b shows that there is no significant relation between the digital literacy with the time to take measures.

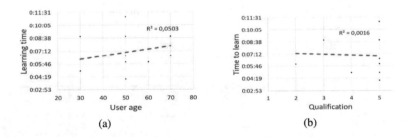

Fig. 6. Learning curve vs age (a) and digital literacy (b).

5   Conclusions and Further Research Lines

This article provides a new eHealth system designed for medical care in nursing homes for the pre-diagnosis diseases. The combination makes the system easy to deploy because of the similarities of existing procedures and the unnecessary skills by the users and optimal for data quality gathering due to the lack of interruptions in the process, as the vital signs selected, body temperature, EDA, pulse oximeter, heart beat rate and blood pressure are hidden in the field for both the resident and the nurse. The App proposed is suitable for taking fast measurements of these vital signs. Data is available online in a SaaS DB accessible from a Web portal and directly by e-Health approved applications via API. The App was designed to be simple, fast to learn, and easy-to-use by non-skilled personnel. Measurements can be taken offline and saved in the local tablet storage capacity until they are uploaded to the online DB, making the App can be used anywhere in the world, even in rural areas. On the other hand, the App is reliable in the sense that when it receives frames are out of range the measures cannot be saved until all sensors provide valid magnitudes.

Possible further research lines are to apply to the clinical DB data mining techniques for pattern recognition for improving the disease diagnosis and treatment. Another line is the generation of an expert system for doctors based on the data models built form the DB. And third foreseen application is centralized monitoring systems improving existing services that cannot react with anticipation.

Acknowledgements. This work is part of the AC16/00061 project: "Design and implementation of a low-cost intelligent system for pre-diagnosis and telecare of infectious diseases in the elderly (SPIDEP)" financed with resources from Instituto de Salud Carlos III (Ministry of Science, Innovation and Universities of Spain), together with the Fundación para la Investigación Biomédica del Hospital Universitario Príncipe de Asturias, Spain, the National Secretariat of Science, Technology and Innovation of Panamá (SENACYT), through the National Research System, Panamá, and all of this within the 2nd Joint Call for Research and Innovation ERANet-LAC within the European Union 7th Framework Programme for Research and Technology Development (FP7) under the ELAC2015/T09-0819 SPIDEP contract.

References

1. Adaji, A., Melin, G.J., Campbell, R.L., Lohse, C.M., Westphal, J.J., Katzelnick, D.J.: Patient-centered medical home membership is associated with decreased hospital admissions for emergency department behavioral health patients. Popul. Health Manag. **21**(3), 172–179 (2018). https://doi.org/10.1089/pop.2016.0189
2. Forni, A.A., van der Meulen, R.: Gartner says worldwide sales of smartphones grew 9 percent in first quarter of 2017 (2017). https://www.gartner.com/en/newsroom/press-releases/2017-05-23-gartner-says-worldwide-sales-of-smartphones-grew-9-percent-in-first-quarter-of-2017
3. Berndt, R., Takenga, M.C., Kuehn, S., Preik, P., Sommer, G., Berndt, S.: SaaS-platform for mobile health applications. In: International Multi-Conference on Systems, Signals & Devices, pp. 1–4 (2012). https://doi.org/10.1109/SSD.2012.6198120
4. Biosignalplux. https://www.biosignalsplux.com/en/
5. Catarinucci, L., de Donno, D., Mainetti, L., Palano, L., Patrono, L., Stefanizzi, M.L., Tarricone, L.: An IoT-aware architecture for smart healthcare systems. IEEE Internet Things J. **2**(6), 515–526 (2015). https://doi.org/10.1109/JIOT.2015.2417684
6. Chatwin, M., Hawkins, G., Panicchia, L., Woods, A., Hanak, A., Lucas, R., Baker, E., Ramhamdany, E., Mann, B., Riley, J., Cowie, M.R., Simonds, A.K.: Randomised crossover trial of telemonitoring in chronic respiratory patients (Tele-CRAFT trial). Thorax **71**(4), 305–311 (2016). https://doi.org/10.1136/thoraxjnl-2015-207045

7. Costa, F.F.: Big data in biomedicine. Drug Discov. Today **19**(4), 433–440 (2014). https://doi.org/10.1016/j.drudis.2013.10.012

8. Fan, L., Buchanan, W., Thummler, C., Lo, O., Khedim, A., Uthmani, O., Lawson, A., Bell, D.: DACAR platform for eHealth services cloud. In: 2011 IEEE 4th International Conference on Cloud Computing, pp. 219–226 (2011). https://doi.org/10.1109/CLOUD.2011.31

9. High, K.P., Bradley, S.F., Gravenstein, S., Mehr, D.R., Quagliarello, V.J., Richards, C., Yoshikawa, T.T.: Clinical practice guideline for the evaluation of fever and infection in older adult residents of long-term care facilities: 2008 update by the Infectious Diseases Society of America. Clin. Infect. Dis. **48**(2), 149–171 (2009). https://doi.org/10.1086/595683

10. Kontis, V., Bennett, J.E., Mathers, C.D., Li, G., Foreman, K., Ezzati, M.: Future life expectancy in 35 industrialised countries: projections with a Bayesian model ensemble. Lancet **389**(10076), 1323–1335 (2017). https://doi.org/10.1016/S0140-6736(16)32381-9

11. Kumar, N.M., Mallick, P.K.: The Internet of Things: insights into the building blocks, component interactions, and architecture layers. Procedia Comput. Sci. **132**, 109–117 (2018). https://doi.org/10.1016/j.procs.2018.05.170. http://www.sciencedirect.com/science/article/pii/S1877050918309049

12. Newman, S.: Building Microservices: Designing Fine-Grained Systems. O'Reilly Media Inc., Boston (2015)

13. Ong, M.K., Romano, P.S., Edgington, S., Aronow, H.U., Auerbach, A.D., Black, J.T., De Marco, T., Escarce, J.J., Evangelista, L.S., Hanna, B., Ganiats, T.G., Greenberg, B.H., Greenfield, S., Kaplan, S.H., Kimchi, A., Liu, H., Lombardo, D., Mangione, C.M., Sadeghi, B., Sadeghi, B., Sarrafzadeh, M., Tong, K., Fonarow, G.C.: Effectiveness of remote patient monitoring after discharge of hospitalized patients with heart failure: the better effectiveness after transition – heart failure (BEAT-HF) randomized clinical trial. JAMA Intern. Med. **176**(3), 310–318 (2016). https://doi.org/10.1001/jamainternmed.2015.7712

14. Shakshuki, E., Reid, M.: Multi-agent system applications in healthcare: current technology and future roadmap. Procedia Comput. Sci. **52**, 252–261 (2015). https://doi.org/10.1016/j.procs.2015.05.071. http://www.sciencedirect.com/science/article/pii/S1877050915008716

15. Simmhan, Y., Ravindra, P., Chaturvedi, S., Hegde, M., Ballamajalu, R.: Towards a data-driven IoT software architecture for smart city utilities. Softw. Pract. Exp. **48**(7), 1390–1416 (2018)

16. Taichman, D.B., Backus, J., Baethge, C., Bauchner, H., de Leeuw, P.W., Drazen, J.M., Fletcher, J., Frizelle, F.A., Groves, T., Haileamlak, A., James, A., Laine, C., Peiperl, L., Pinborg, A., Sahni, P., Wu, S.: Sharing clinical trial data: a proposal from the International Committee of Medical Journal Editors (2016). https://doi.org/10.4314/ejhs.v26i1.2

17. Uqaili, A.: Smartphone use among young doctors and their impact on patients of Liaquat University Hospital Jamshoro. Imperial J. Interdiscip. Res. (IJIR) **3**, 161–164 (2017)

18. Ventola, C.L.: Mobile devices and apps for health care professionals: uses and benefits. P & T Peer-Rev. J. Formul. Manag. **39**(5), 356–364 (2014). https://www.ncbi.nlm.nih.gov/pubmed/24883008, https://www.ncbi.nlm.nih.gov/pmc/articles/PMC4029126/

19. Vilaplana, J., Solsona, F., Abella, F., Filgueira, R., Rius, J.: The cloud paradigm applied to e-Health. BMC Med. Inform. Decis. Mak. **13**(1), 35 (2013). https://doi.org/10.1186/1472-6947-13-35

20. Wu, P., Nam, M.Y., Choi, J., Kirlik, A., Sha, L., Berlin, R.B.J.: Supporting emergency medical care teams with an integrated status display providing real-time access to medical best practices, workflow tracking, and patient data. J. Med. Syst. **41**(12), 186 (2017). https://doi.org/10.1007/s10916-017-0829-x

21. Zhang, R., Liu, L.: Security models and requirements for healthcare application clouds. In: 2010 IEEE 3rd International Conference on Cloud Computing, pp. 268–275 (2010). https://doi.org/10.1109/CLOUD.2010.62

Wearable Tracking Modules Based on Magnetic Fields

Michael Martinez[1], José Baca[1(✉)], Juan Martinez[2], and Mason Myers[1]

[1] Department of Engineering, Texas A&M University-Corpus Christi,
6300 Ocean Dr., Corpus Christi, TX, USA
mmartinez102@islander.tamucc.edu, jose.baca@tamucc.edu
[2] Department of Computer Science, Texas A&M University-Corpus Christi,
6300 Ocean Dr., Corpus Christi, TX, USA

Abstract. This work presents the development of a novel methodology in magnetic tracking (localization). Our approach is based on using the Jacobian matrix of the magnetic flux density from a dipole magnet. Magnetometers that will sense the magnetic flux density will thus be able to determine the position and orientation of a magnetic dipole source. This project leverages the development of an experimental platform that could be used within the health care domain. The use of this platform could be extended to different applications such as in space for monitoring astronauts during the execution of strength training and aerobic exercises, senior care for evaluating rehabilitation therapies, the assessment of human body movements of stroke survivors, as well as, other patients with sensorimotor disorders.

1   Introduction

Magnetic tracking (localization) systems involve the use of magnetic sensing to determine the position and orientation of a magnetic source. Magnetic tracking systems enjoy benefits that some tracking systems do not, such as the ability to track without a line of sight. This feature makes magnetic tracking systems an attractive localization technique. For instance, within the medical field, there is a system know as magnetically actuated capsule endoscopy (MACE) [7,11], that utilizes a magnetic dipole source to power an untethered robot. In this example, a magnetic tracking system could be used to obtain the position and orientation of the dipole source. In applications such as augmented reality/virtual reality, magnetic tracking systems could be utilized as a replacement for the computer vision techniques, which generally suffer from occlusion [2]. Other applications include the use of magnetic tracking systems to localize physical modules connected by magnetic sources [6]. Since permanent magnets are already being utilized to provide connections between modules, the tracking of the modules' position/orientation could also be estimated.

Magnetic tracking systems usually follow a specific recipe in terms of their implementation. The process is usually defined by three steps: (1) Measure magnetic flux densities, (2) Process the measured fields, (3) Perform algorithm(s) on

A. Martínez et al. (Eds.): LACAR 2019, LNNS 112, pp. 214–223, 2020.
https://doi.org/10.1007/978-3-030-40309-6_21

fields to calculate the position and orientation of the magnetic source. Research on the technique usually differs in the magnetic source type, the sensing equipment, the processing technique of the measured fields, and the algorithms being implemented to perform the computations. For instance, an array of scalar magnetometers has been used as the magnetic sensing equipment in conjunction with a particle-swarm optimization algorithm to calculate the position of the magnetic source in real time [4]. In comparison, researchers primarily use triple-axis magnetometers due their ability to contain more information than scalar magnetometers (such as direction) [2,3,5,12–14]. In some cases, single axis magnetometers are utilized instead [9,11] or in combination with triple axis type [10]. Triple-axis Hall effect sensors were utilized here to increase the amount of information per sensor.

Other type of research focuses on the relationship between the position/orientation errors obtained from a magnetic tracking system to external variables and calculate limitations of such systems [13]. Modifications to the magnetic source are also done in an attempt to measure full six DOF (three dimensional position and orientation) [12], to better fit the magnetic flux density model from a magnetic source [8], or eliminate the Earth's background relatively static field [2]. The position tracking errors are usually in the same range. A position/orientation error of 2.1 ± 0.8 mm and $6.7 \pm 4.3°$ was measured when utilizing a second order differentiation technique [11]. However, the error has been shown to vary as the magnetic source is positioned increasingly farther away from the magnetic sensing equipment [13].

This paper focuses on the derivation and implementation of a novel magnetic tracking system that utilizes the Jacobian matrix of the measured magnetic flux densities from a magnetic dipole source to determine the dipoles position and orientation. This novel technique is compared to a standard method of determining the position and orientation of a magnetic source using magnetic flux densities alone. An existing Jacobian technique has been implemented before [7], however, as will be discussed later, the spatial derivatives in the magnetic flux density are measured by a single moving capsule (containing 6 orthogonal magnetometers and an IMU) at two different periods of time whereas the Jacobian method presented here measures such spatial derivatives in the magnetic flux density simultaneously. The former method thus requires absolute localization data from the on-board IMU in conjunction with the 6 orthogonal magnetometers. Further discussion over the differences between this method [7] and our approach, and the implications are shown later in this work.

2   Derivation of Jacobian Matrix Method

Since the main objective in a magnetic tracking system is to find the position and orientation of a magnetic dipole source, one must find where the position and orientation information of a magnetic dipole is encoded. As an example, Fig. 1 illustrates the implementation of this concept and how it attempts to solve it.

We therefore define a position vector **r** which represents the displacement from the magnetic dipole source to the location of the sensor(s). If the position

Fig. 1. Wearable magnetic strap module concept. The main technical challenges include (1) real-time estimation of angular position of the joint; (2) low-complexity system design in both software and hardware via modularity concept; (3) compact size to fit in portable modules; and (4) accuracy in detecting slight changes in magnetic flux density from magnetic source.

of the sensor(s) is known, then this position vector will encode the location of the magnetic dipole source relative to the sensor(s), which is the position information of the dipole source we are looking for. To obtain the orientation of a dipole source, one can look at the magnetic moment vector ($\mathbf{m}$) of the dipole source. The magnetic moment vector points in the direction in which the dipole source is oriented. Thus, by finding the magnetic moment vector, one can calculate the direction of the magnetic dipole source, and therefore the orientation. In conclusion, in order to track the position and orientation of a magnetic dipole source, the position and magnetic moment vectors ($\mathbf{r}$ and $\mathbf{m}$, respectively) must be obtained mathematically.

The approach taken here does not use numerical methods to solve for $\mathbf{r}$ and $\mathbf{m}$, instead, it collects spatial derivatives of magnetic flux density which yields to that of the inverse of the Jacobian matrix $\mathbf{P}$

$$\mathbf{P} = \begin{bmatrix} \partial_x B_x & \partial_y B_x & \partial_z B_x \\ \partial_x B_y & \partial_y B_y & \partial_z B_y \\ \partial_x B_z & \partial_y B_z & \partial_z B_z \end{bmatrix} \tag{1}$$

It can be used to obtain the position vector $\mathbf{r}$.

$$\mathbf{r} = (\mathbf{P}^{-1}\mathbf{B}) \tag{2}$$

Since the exact partial spatial partial derivative terms of $\mathbf{B}$ (magnetic flux density of magnetic dipole) can not be determined, they must be approximated

$$\mathbf{B} = \begin{bmatrix} (a-x)\partial_x B_x + (b-y)\partial_y B_x + (c-z)\partial_z B_x \\ (a-x)\partial_x B_y + (b-y)\partial_y B_y + (c-z)\partial_z B_y \\ (a-x)\partial_x B_z + (b-y)\partial_y B_z + (c-z)\partial_z B_z \end{bmatrix} \tag{3}$$

$$\begin{bmatrix} B_x \\ B_y \\ B_z \end{bmatrix} = \begin{bmatrix} \partial_x B_x & \partial_y B_x & \partial_z B_x \\ \partial_x B_y & \partial_y B_y & \partial_z B_y \\ \partial_x B_z & \partial_y B_z & \partial_z B_z \end{bmatrix} \begin{bmatrix} a - x \\ b - y \\ c - z \end{bmatrix} \tag{4}$$

Thus, the position and magnetic moment vectors are decoupled and are found to be given by the following equations:

$$\mathbf{r} = (\mathbf{P}^{-1}\mathbf{B}) \tag{5}$$

$$\mathbf{m} = \frac{4\pi}{\mu_o} r^5 (\mathbf{T}^{-1}\mathbf{B}) \tag{6}$$

where $\mathbf{T}$ is defined as

$$\mathbf{T} = \begin{bmatrix} (3x^2 - r^2) & 3xy & 3xz \\ 3yx & (3y^2 - r^2) & 3yz \\ 3zx & 3zy & (3z^2 - r^2) \end{bmatrix} \tag{7}$$

With the now calculated position and magnetic moment vectors, the position of the magnetic dipole source relative to the sensor(s) and its orientation (a total of five DOF) can be found.

3   Implementation of Jacobian Approximation

To make use of Eqs. (5) and (6), approximations for the Jacobian matrix in Eq. (1) are necessary. This is because the actual spatial derivatives (single point slopes) of the magnetic flux density cannot be determined using Hall-effect sensors alone. For each partial derivative of the magnetic flux density components, the following approximations can be made:

$$\partial_x B_x \approx \frac{B_x(x + \delta x, y, z) - B_x(x, y, z)}{\delta x} \tag{8}$$

$$\partial_y B_x \approx \frac{B_x(x, y + \delta y, z) - B_x(x, y, z)}{\delta y} \tag{9}$$

$$\partial_z B_x \approx \frac{B_x(x, y, z + \delta z) - B_x(x, y, z)}{\delta z} \tag{10}$$

$$\partial_x B_y \approx \frac{B_y(x + \delta x, y, z) - B_y(x, y, z)}{\delta x} \tag{11}$$

$$\partial_y B_y \approx \frac{B_y(x, y + \delta y, z) - B_y(x, y, z)}{\delta y} \tag{12}$$

$$\partial_z B_y \approx \frac{B_y(x, y, z + \delta z) - B_y(x, y, z)}{\delta z} \tag{13}$$

$$\partial_x B_z \approx \frac{B_z(x + \delta x, y, z) - B_z(x, y, z)}{\delta x} \tag{14}$$

$$\partial_y B_z \approx \frac{B_z(x, y + \delta y, z) - B_z(x, y, z)}{\delta y} \tag{15}$$

$$\partial_z B_z \approx \frac{B_z(x, y, z + \delta z) - B_z(x, y, z)}{\delta z} \tag{16}$$

With these approximations in mind, the sensor array necessary to calculate the Jacobian matrix approximations would be as shown in Fig. 2.

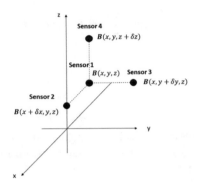

Fig. 2. Arrangement of triple-axis sensors for implementing the Jacobian matrix method.

As seen in Fig. 2, four triple-axis sensors are required to implement the equations. As observed in Eqs. (8–16), a point at x, y, and z (with no displacements in either axis) measuring the magnetic flux density is required. This requirement is represented by Sensor 1 in Fig. 2. Sensors 2–4 represent the displacements along the x, y, and z axes, respectively. For each point, all three components of the magnetic flux density are readily available for computation. With this sensor configuration, it is crucial that the displacements between sensors are made as small as possible. Specifically, they must be significantly smaller than the tracking distances between the sensor array and magnetic dipole source. Furthermore, the sensors must be sensitive enough to detect differences between the displaced sensors and Sensor 1.

Regarding the approximation scheme, the resulting Jacobian matrix in Eq. (1) for calculating the position of the dipole source can then be shown to be approximated by substitution of Eqs. (8–16). The result is the valid use of Eqs. (5) and (6) with the now redefined matrix. The equations that calculate the position vector are independent of the medium properties and the magnetic dipole moment vector. Once the position vector is calculated, the magnetic dipole moment can be determined with the use of the calculated position vector. Five DOF of the dipole source can then be obtained.

4   Wearable Magnetic Tracking Module

To validate the approach, it is necessary to implement the strategy into a physical prototype. We developed a wearable module that can be worn by the user to monitor movements based on joint rotations of the human body, for instance, elbow rotations as shown in Fig. 3. This modular system could provide the user with information related to the training and performance parameters such as, range of motion, timing, repetitions, etc. [1]. The sensor system senses the magnetic fields generated by the dipole magnet. Data is sent to microcontroller where equations described in Sect. 2 continuously calculate elbow's angle (θ). In this work, the system does not use any type of filter and calculations are based on raw data.

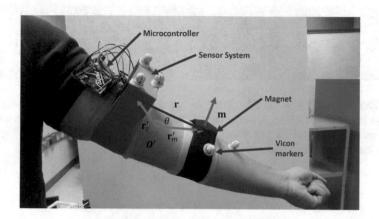

Fig. 3. A magnetic tracking module is composed by a sensor system/microcontroller (mounted on a blue strap), and a dipole magnet (mounted on a black strap). This module could be mounted to different joints of the human body, e.g., elbow, wrist, knee, ankle, etc. In this scenario, the system will be monitoring the rotation of the elbow. The sensor system is formed by the proper arrangement of sensors previously discussed. For comparison purposes during the experiments, the elbow's angle θ will be calculated by both, magnetic tracking module and Vicon system.

To test the prototype, a magnetic tracking module was mounted at the elbow of a user. By using the calculated displacement vector of the magnet relative to the sensor system, the angle formed between the position vectors of both the sensor system and magnet relative to a local origin O' could be calculated. As shown in Fig. 3, the tracking system is characterized by blue and black straps. The blue strap contains the sensor arrangement, while the black strap contains the dipole cylindrical magnet. m is the magnetic moment vector of the magnet, r is the displacement vector of the magnet relative to the sensor system, r'_m is the position of the magnet relative to the local origin, r'_s is the position of the sensor system relative to the local origin, and θ is the angle between the position vectors r'_s and r'_m.

5   Experimental Results

In order to evaluate the performance of the approach, two main experiments were performed. The first experiment consists of positioning the elbow at different constant angles and comparing results between our proposed approach and those true angles given by the Vicon system. The second experiment consists of constantly varying the elbow's angle at low and high frequency, as shown in Fig. 4.

Fig. 4. Experimental test bed. User wears a magnetic tracking module at the elbow and executes joint rotations. Module calculates the elbow's angle during the two types of experiments, i.e., positioning elbow at different constant angles, and varying elbow's angle at low and high frequency.

5.1   Positioning Elbow at Different Constant Angles

In this experiment, the user tries to keep the elbow at a constant angle. Four different runs were executed for a total of four constant elbow's angles. As shown in Fig. 5, one can notice quantitative errors between calculated angles and true angles (some errors as large as $100°$). However, general features of the calculated angle closely resemble to those of the Vicon system (true angle) during the entire run (sample number = timestamp). Even small angular variations measured by the Vicon system, are also calculated by our approach. Implications of this observation will be discussed later in conclusions.

5.2   Varying Elbow's Angle at Low and High Frequency

In this experiment, the user rotates elbow to generate angular motion at different speeds. Two runs were performed at varying frequencies, first test is performed at a low frequency (user rotates elbow back and forth, repetitively) and the second test is performed at high frequency (faster repetitions). Experimental results display that our approach finds challenging to calculate the angles at varying speeds, as shown in Fig. 6. Unlike in previous experiment, general features of the calculated angles hardly mimic that of the Vicon data as the frequency of the repetition increases. One of the reasons of such problem is the induced voltage to the sensors, which causes noise. The faster the elbow's angle changes over time, the larger the noise levels in the sensor system. Another reason is that this approach is using raw data from the sensors to calculate the angle, therefore, filtering techniques will be required to better estimate angular positions during joint rotations.

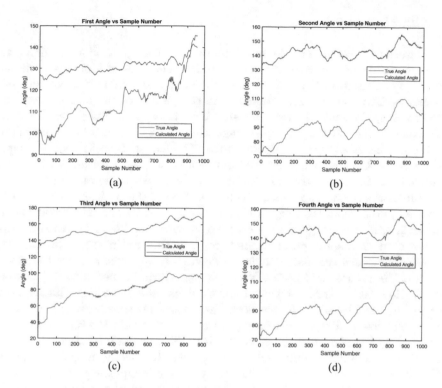

(a)

(b)

(c)

(d)

Fig. 5. Experimental data from tests in which user tries to keep four constant angles over a period of time. By keeping angle variation to a minimum, the calculated angle follows closely to the true angle. Movement's patterns over the entire test are properly followed by our approach, however, there are quantitative errors between calculated and true angles that have to be considered. Each graph represents a set of angles randomly set by user and slowly changed over time.

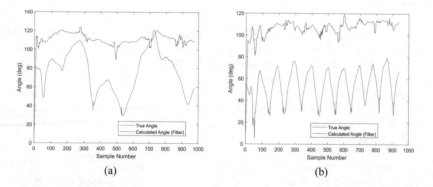

(a)

(b)

Fig. 6. Experimental results from tests in which user rotates elbow over a period of time. Approach finds challenging to calculate angular rotations during repetitions at low and high frequency. Filtering techniques will be required to address induced voltage within sensor system.

6   Conclusions

In this work we have presented a novel magnetic tracking technique utilizing the Jacobian matrix of the measured magnetic flux densities of a magnetic dipole. The main contribution of this work is in the ability to create portable magnetic tracking systems through reduction in the complexity of the position and orientation computations. The main difference of this approach with respect other techniques, is the creation of a new magnetic tracking methodology that allows for the tracking of a magnetic dipole source in 5 DOF without the use of numerical methods. Usually, the magnetic moment vector and position vector must be solved for simultaneously. The technique also serves to decouple to position vector from the magnetic moment vector. This way, the position vector can be evaluated purely from the magnetic flux density and its spatial derivatives. In this work, while the magnetic moment vector equation could be calculated, it was not used for angular calculations here. Experimental results show that the presented approach calculates angles in such as a way that mirrors the movement's patterns measured by the Vicon system during experiment 1 (constant angles). However, it is important to mention that there is a quantitative error that should be addressed in future work. From results within experiment 2 (varying angle at low and high frequency), it was observed that the induced voltage generated during the repetitions, does affect the calculations. However, quantitative error is smaller compared to experiment 1. Usually magnetic tracking suffers from four major sources of error: real magnetic dipole source approximation, background noise from induced electrical currents (if the magnetic dipole source is moving, as shown in the last two experimental trials), low sensor resolution, background noise fields (Earth's magnetic field) and the dipole equation approximation. The Jacobian matrix method is affected by these factors, as well as, due to the approximation of the first order spatial partial derivatives.

References

1. Baca, J., Ambati, M.S., Dasgupta, P., Mukherjee, M.: A modular robotic system for assessment and exercise of human movement. In: Advances in Automation and Robotics Research in Latin America, pp. 61–70. Springer (2017)
2. Chen, K.Y., Patel, S.N., Keller, S.: Finexus. In: Proceedings of the 2016 CHI Conference on Human Factors in Computing Systems, CHI 2016. ACM Press (2016)
3. Dai, H., Yang, W., Xia, X., Su, S., Ma, K.: A three-axis magnetic sensor array system for permanent magnet tracking. In: 2016 IEEE International Conference on Multisensor Fusion and Integration for Intelligent Systems (MFI). IEEE (2016)
4. Fan, L., Kang, C., Zhang, X., Zheng, Q., Wang, M.: An efficient method for tracking a magnetic target using scalar magnetometer array. SpringerPlus 5, 1–10 (2016)
5. Hu, C., Meng, M.H., Mandal, M.: A linear algorithm for tracing magnet position and orientation by using three-axis magnetic sensors. IEEE Trans. Magn. 43(12), 4096–4101 (2007)
6. Nagy, Z., Abbott, J.J., Nelson, B.J.: The magnetic self-aligning hermaphroditic connector a scalable approach for modular microrobots. In: 2007 IEEE/ASME International Conference on Advanced Intelligent Mechatronics. IEEE (2007)

7. Natali, C.D., Beccani, M., Simaan, N., Valdastri, P.: Jacobian-based iterative method for magnetic localization in robotic capsule endoscopy. IEEE Trans. Robot. **32**(2), 327–338 (2016)
8. Raab, F., Blood, E., Steiner, T., Jones, H.: Magnetic position and orientation tracking system. IEEE Trans. Aerosp. Electron. Syst. **AES-15**(5), 709–718 (1979)
9. Schlageter, V., Besse, P.A., Popovic, R., Kucera, P.: Tracking system with five degrees of freedom using a 2D-array of Hall sensors and a permanent magnet. Sens. Actuators, A **92**(1–3), 37–42 (2001)
10. Sherman, J.T., Lubkert, J.K., Popovic, R.S., DiSilvestro, M.R.: Characterization of a novel magnetic tracking system. IEEE Trans. Magn. **43**(6), 2725–2727 (2007)
11. Son, D., Yim, S., Sitti, M.: A 5-D localization method for a magnetically manipulated untethered robot using a 2-D array of Hall-effect sensors. IEEE/ASME Trans. Mechatron. **21**(2), 708–716 (2016)
12. Song, S., Qiu, X., Liu, W., Meng, M.Q.H.: An improved 6-D pose detection method based on opposing-magnet pair system and constraint multiple magnets tracking algorithm. IEEE Sens. J. **17**(20), 6752–6759 (2017)
13. Su, S., Yang, W., Dai, H., Xia, X., Lin, M., Sun, B., Hu, C.: Investigation of the relationship between tracking accuracy and tracking distance of a novel magnetic tracking system. IEEE Sens. J. **17**(15), 4928–4937 (2017)
14. Wahlstrom, N., Gustafsson, F.: Magnetometer modeling and validation for tracking metallic targets. IEEE Trans. Sig. Process. **62**(3), 545–556 (2014)

Design of an Inclusive Early Warning System. Case of Basin of Pacora River, Panama

Ignacio Chang[1](✉), Antony García[1], and Ernesto García[2]

[1] Facultad de Ingeniería Eléctrica, Universidad Tecnológica de Panamá, Panama City, Panama
ignacio.chang@utp.ac.pa
[2] Facultad de Ingeniería de Sistemas Computacionales, Universidad Tecnológica de Panamá, Panama City, Panama

Abstract. An early warning system (EWS) reduces the loss of lives and decreases the economic and material impact on vulnerable populations. The work presented in this article consists of a prototype system designed in accordance with the Disability Policy of the Republic of Panama such as the elimination of all forms of exclusion and discrimination for people with disabilities, elimination of physical barriers for the full participation of people with disabilities or social protection to improve the quality of life and greater independence of the population with disabilities and their families. This innovative system links the necessary elements for early and timely warning, and includes the role of the human element of the system and risk management. In our case, the inclusive EWS implied a set of activities that had as goals the generation of a hydraulic model of the river, a sensorial network for the detection of the warning and a communication system for the diffusion of the warning that includes people with visual or auditory disabilities. The inclusive EWS was divided into several sub-systems. One of these was related to the measurement stations, called MOSU, another with the detection of the warning and the diffusion mechanism named SUDI and finally, the device that warns the person the warning itself known as SUIN. The system works in real time through the telemetric interconnection of a sensor network connected to a wireless network that registers the levels of the Pacora river and that these, together with a calibrated hydraulic model of the river, allows for the precise determination of moments in which there is a risk of flood. These sensors monitor the river in its upper, middle and lower part, thus allowing for an automated and efficient Early Warning System.

Keywords: Inclusive alert system · LORA network · Sensory network · Communication system

1 Introduction

An Early Warning System (EWS) reduces the loss of life and decreases the economic and material impact on the affected vulnerable populations. Its effectiveness lies fundamentally in: (a) knowledge of the existence of risks, (b) the

A. Martínez et al. (Eds.): LACAR 2019, LNNS 112, pp. 224–235, 2020.
https://doi.org/10.1007/978-3-030-40309-6_22

active participation of communities and (c) an institutional commitment that involves education as an indispensable factor for citizen awareness and dissemination [6]. Therefore, the development of projects that provide an opportunity for the population to protect themselves from disasters and safeguard the ecosystem, with the consequent improvement in the quality of life of the population, is imperative. In the literature we find a set of EWSs in various parts of the world, for example in [10–14]. As in these cases, each one describes the particularities and benefits of the proposed system but most of them do not include people with any type of disability. Due to the importance of these systems on people and goods, as described in [1–3, 5, 7, 8], several academic and research units came together to design and install a prototype EWS that would include people with visual or auditory disability. People with disabilities were included because according to the last Population Census of 2010 the percentage of disability was 2.9%; Of these, 22% correspond to people with visual disabilities and 15.6% to people with hearing disabilities and it is increasing [9].

One of the 35 priority basins in the country with a lot of impact is the Pacora river basin, which, over time, the hand of man has caused changes in its environment and land uses, thus reducing the potential retention of the basin at the time of extreme rainfall such as the phenomenon the purest occurred a few years ago.

This basin is characterized by:

- Historical data,
- Growth without any planning,
- An off line monitoring system installed but unsupervised,
- The existence of contacts with community members (community leaders) Being close to the facilities of the Technological University of Panama (a research center where the authors of this article belong), which reduces the costs of mobilization,
- Use of land not corresponding to a functional land use plan,
- A sustained increase in the degree of vulnerability of the population and the ecosystem of the basin as a consequence of the constant flooding and the contamination of its waters, mainly due to the violation of compliance with environmental parameters, which may lead to loss of human lives, water resources and biodiversity.

For these reasons this basin was chosen as a testing and experimentation center for the development of a prototype EWS. In addition, the Municipality of Panama in its Districtal Strategic Plan [?], which includes the change of forest cover, characterizes this basin and proposes development alternatives that support the decision to consider this basin to implement a EWS prototype. An innovative contribution that we incorporate into this design is the development of electronic assistance devices for people with hearing or visual disabilities considering the fact of the rapid growth of this segment of the population in the country and due to its greater vulnerability as people with disabilities. In this way, Sect. 2 presents the design strategy that involves the telemetric system along the Pacora River and some of its tributaries, with level sensors for the real-time capture of the river levels, whose data feeds a mathematical model for the hydraulic simulation and prediction of flood risk levels in areas identified as vulnerable. Section 3 describes the components of the system from the designed subsystems, while in the fourth section describes the characteristics of the system and some considerations about the technology used. The fifth is a discussion of the results obtained and, finally, the conclusions.

2 Design Strategies

Instead of simply listing headings of different levels we recommend to let every heading be followed by at least a short passage of text. Further on please use the LATEX automatism for all your cross-references and citations. The work presented in this article consists of a prototype system designed in accordance with the Disability Policy of the Republic of Panama [4] including: elimination of all forms of exclusion and discrimination for people with disabilities, elimination of barriers for the full participation of people with disabilities or social protection to improve the quality of life and greater independence of the population with disabilities and their families. This novel system links the elements necessary for early and timely warning, and includes the role of the human element of the system and risk management.

To achieve this, we proceeded to organize a work team in several subgroups, with specific and well-defined tasks. One of the subgroups of work was dedicated to the hydraulic analysis of the river, another related to the instrumentation and communication platform, and the last one for a socio-economic analysis of the most vulnerable area.

2.1 Hydraulic Analysis

The following activities were carried out: (a) topographic surveys and geodesic moorings, (b) determination of geomorphological and morphogeometric parameters of the river channel, (c) determination of the HEC-RAS floodplains and (d) calibrating the model and determining warning levels.

This task involved a series of mobilizations to the basin to perform measurements, clearing, transfer of specialized equipment, collection of location data and the use of the HEC-RAS program to calibrate the model and determine warning levels. Figure 1 shows part of those measurements and devices used.

Figure 2 shows (a) the topographic survey of the Pacora River carried out by the Center for Hydraulic and Hydrotechnical Research (CIHH) of the Technological University of Panama (UTP). When applying the HEC-RAS program, the flood plains at 50 years were obtained (b) as observed.

With the historical data, plus those collected during 8 months off-line, an analysis was carried out that determined that a head of water takes 40 min to reach the lower part of the river, this area coincides with the area of greatest concentration of the population. This time makes it easier to get people out of the area of possible flooding in the event of the alert.

Fig. 1. Views of gaugings and surveys

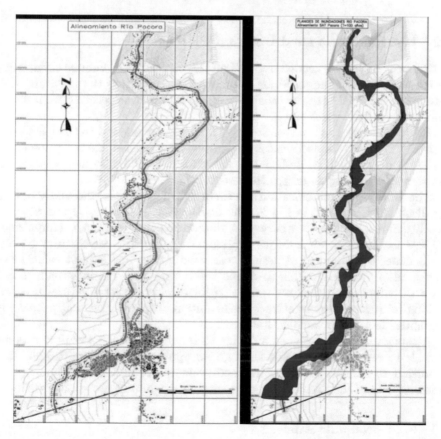

Fig. 2. Alignment of the Pacora river and floodplains

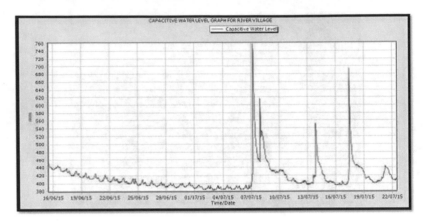

Fig. 3. Sample of measurements

2.2   Instrumentation and Diffusion Platform

In this case, tests were carried out in the UTP to verify the wireless connectivity for the transmission and reception of the warning, as well as the structure of the database associated with the reliability of the information received from distant points transmitted wirelessly.

Figure 3 shows a run of river levels. The peaks correspond to the flash floods in the river. Figure 4 shows what is the supply, transmission and reception circuit of the warning signal, all designed in the laboratory of Embedded Systems and Assistance Technologies, except the antenna.

The prototype communication system designed for people with visual or hearing disabilities is shown in Fig. 5. The device that issues the alert is not completely defined, however, in a survey conducted on people with visual impairment resulted in the choice of a device similar to an electronic keychain or an electronic cane. It is planned to design both alternatives and submit them to the consideration of these people to define the device they will eventually use. It is worth emphasizing that we want this final device to have several uses, not striking so that it does not attract the attention of thieves and that people with disabilities feel comfortable with its use.

Although we have designed a device to communicate the warning, it is also possible to receive the warning through an application in a smartphone. It is important to indicate that the designed system is of universal technology, the type and uses of the device for people without disabilities are already defined.

2.3   Socioeconomic Analysis

To carry out the socio-economic analysis, it was decided to generate a survey (see Fig. 5). It was decided to apply it only in the most vulnerable areas, that is, in the communities settled on the banks of the river.

Fig. 4. Circuitry for feeding, receiving and transmitting data

Fig. 5. Area visited in red

From the analysis of the surveys, one of the most significant problems of the surveyed was the waste management. Another finding was that in the five communities only five people among the parents did not know how to read or write, of them four were fathers and one mother. This fact shows a low index of "no schooling". In the economic sphere, the Universal Scholarship program is the one that most beneficial for the communities studied, and in general it benefits 63% of the families. The economic activities carried out are predominately agriculture (22.0%) and Poultry (18.0%), in the majority of households is perform for self-consumption.

3   System Components

For a better understanding of the system, it has been divided into several subsystems. One of these is related to the measurement stations, another with the detection of the warning and the diffusion mechanism and finally, the device that warns the person the warning itself. Let's see each one of them.

3.1   Monitoring Subsystems (MOSU)

Consists of three measurement points. The first in the upper part of the river, the second in the middle part and the last in the lower part. Figures 6 and 7 show these points located in the upper, middle and lower part of the river. The construction of the measuring stations (limnigraphics) were made with conventional materials, such as: concrete and PVC accessories. The proposed limnigraphic stations are of low cost and as resistant as possible to the strong currents produced during floods. Figure 8 shows the design of the station and its location on the ground.

The sensors capture the hydrostatic level by means of capacitance and, based on the calibration of the sensor, transforms said capacitance to water level. Unlike other sensors that measure the hydrostatic pressure and whose measurements must be adjusted in function to the variations in the atmospheric pressure.

Fig. 6. Construction of the limnigraphic station in the upper part of the river

Fig. 7. Stations located in the middle and lower parts of the river

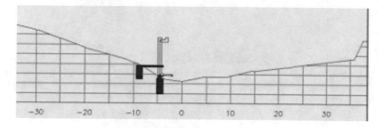

Fig. 8. Design of the limnigraphic stations

3.2   Subsystem of Diffusion (SUDI)

It is composed of different types of nodes that are found along the river bank, in flood influence zones and in command centers. These nodes form the communication network that supports the transmission of messages to alert people to flood events. In Fig. 9, points 1, 2, 3, (measurement stations), the center (unit of detection of the warning and where data is analyzed) and SINAPROC which means Civil Protection System (entity that issues the alert) form the backbone of data communication network. This part of the network works as a bidirectional communication bus that transmits the data and the warning to SINAPROC, which, in turn, disseminates the warning to the people who are within the areas of incidence for the evacuation if necessary. In addition, points 2 and 3, form the main point of communication from where the warning messages are transmitted to the different safe points (shelters) and from there to the devices that people have in the different areas of incidence. Remember, it is universal design.

This network incorporates different types of communication technology depending on the conditions and limitations of the place. Between the points 1,2 and 3, the refuges and the devices of the people LoRaWAN is used since they are locations where there is no cellular coverage. In this sense LoRAWAN is a technology that allows us to transmit at a greater distance but at a lower transmission speed; for this scenario it is sufficient since the message transmitted allows the person's device to vibrate to indicate the warning.

3.3   Inclusive Subsystem (SUIN)

Is a device whose central unit is an embedded system that produces a vibration at the moment of receiving the warning signal, warning the people who are in the area of the incident by flood or by flash flood and has a record of people including those with visual or hearing impairments. In this way, only in the case of an alert will the people, with or without registered disabilities, become visible to the authorities in charge of managing the disaster and their geographical location. At all times if they can know the number of people who are in the area of possible flood (revisar el parrafo en español y editarlo de forma que haga sentido).

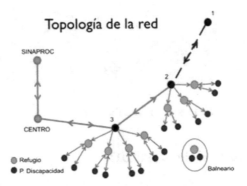

Fig. 9. Transmission and reception network

In Fig. 9 the people monitored in the possible area of disaster appear in red color. Also, there is an isolated section selected by the local authorities identified as a bath, this section is only for people with visual or hearing disabilities to enjoy the river and in case of warning the device takes them autonomously to the shelter, as it happens if your residence is in the area of the flood. This situation is observed in Fig. 10.

4   System Characteristics

The system will work in real time through the telemetric interconnection of a sensor network connected to a wireless network that registers the levels of the Pacora River (see Fig. 11) and that these, together with a calibrated hydraulic model of the river, can allow precise determination of the moments where there is

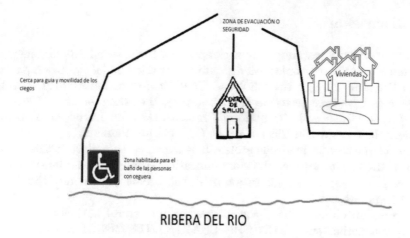

Fig. 10. Communication system for people with visual or hearing impairment

Fig. 11. Measuring stations installed

a risk of flood. These sensors will monitor the river mainly in its upper, middle and lower parts, thus allowing for an automated and efficient Early Warning System.

The proposed design incorporates people with visual or hearing disabilities, the warning signal reaches people within the disaster area, local authorities and civil defense organizations. Applies tools such as electronics, radio frequency, LORA network and a wireless network. It uses an energy system based on photovoltaic energy.

5   Discussion

The designed early warning system complies with the standards for the protection and assistance of people with disabilities obtained under the accompaniment of the Program through the RET and UNICEF Education in 2019. The three subsystems have been tested in the laboratory, the measurements of level were carried out off-line and the hydraulic model also used historical data in the Empresa de Transmisión Eléctrica S.A. (ETESA) of Panama.

One of the sensors has been stolen. It is expected to replace it this year and perform the on-site tests of the communication system, as well as the start-up of three pluviographs. A registration of all the people who acquire the device is planned via internet.

The design obtained has been interesting for several national and international organizations such as UNICEF, USAID/OFDA, COSUDE, and in Panama AMUPA, SENADIS, SINAPROC, Mayor of Panama, Ministry of Education.

6   Conclusions

An early warning system has been designed that has as an innovation that includes people with visual or hearing disabilities. Likewise, there is a laboratory with capacity for the design, elaboration of the printed circuit, 3D printing and development of the prototype system ready to be replicated in other priority basins of the country.

The system designed has been liked by many national and international agencies for its low cost, its operation in real time and sending the alert only to people who are within the possible area of disaster.

Acknowledgement. We thank the Central University Council of Central America (CSUCA), the Swiss collaboration (COSUDE), the regional program of technical assistance for the integral management of disaster risk of USAID/OFDA and the Association of Municipalities of Panama (AMUPA) for the support and accompaniment provided in the development of this project.

References

1. Resolución 54/219. decenio internacional para la reducción de los desastres naturales: Nuevas disposiciones. asamblea general. a/res/54/219. http://eird.org/fulltext/GA-resolution/a-res-54-219-spa.pdf
2. Marco de acción de hyogo para 2005–2015. aumento de la resiliencia de las naciones y las comunidades ante los desastres. conferencia mundial sobre la reducción de los desastres (2005). http://www.eird.org/cdmah/contenido/hyogo-framework-spanish.pdf
3. Noticias onu (19 de enero de 2005). onu lanza plan global de sistemas de alerta temprana (2005). https://news.un.org/es/story/2005/01/1049211

4. Política de discapacidad de la república de panamá. comisión para la formulación de la política de discapacidad en panamá (2009). https://www.senacyt.gob.pa/wp-content/uploads/2017/04/Politica-de-Discapacidad-de-la-Rep%C3%BAblica-de-Panam%C3%A1.pdf

5. Década internacional para la reducción de los desastres naturales, de 1990 hasta el año 2000. 85a (2011). https://undocs.org/es/A/RES/44/236

6. Manual sistema de alerta temprana-unesco. panamá (2011). http://www.unesco.org/new/fileadmin/MULTIMEDIA/FIELD/San-Jose/pdf/Panama%20MANUAL%20INFORMATIVO.pdf

7. Manual sobre sistema de alerta temprana (2011). https://plandistritalpanama.com/wp-content/uploads/2019/01/PROD3_PED_tomo-1.pdf

8. Marco de sendai para la reducción del riesgo de desastres 2015–2030 (2015). https://www.unisdr.org/files/43291_spanishsendaiframeworkfordisasterri.pdf

9. Contraloría general de la república de panamá. instituto nacional de estadística y censo (2019). http://www.contraloria.gob.pa/inec/Publicaciones/subcategoria.aspx?ID_CATEGORIA=13&ID_SUBCATEGORIA=59&ID_IDIOMA=1

10. Alfieri, L., Salamon, P., Pappenberger, F., Wetterhall, F., Thielen, J.: Operational early warning systems for water-related hazards in Europe. Environ. Sci. Policy **21**, 35–49 (2012). https://doi.org/10.1016/j.envsci.2012.01.008

11. Balis, B., Bubak, M., Harezlak, D., Nowakowski, P., Pawlik, M., Wilk, B.: Towards an operational database for real-time environmental monitoring and early warning systems. Procedia Comput. Sci. **108**, 2250–2259 (2017). https://doi.org/10.1016/j.procs.2017.05.193

12. Choy, S., Handmer, J., Whittaker, J., Shinohara, Y., Hatori, T., Kohtake, N.: Application of satellite navigation system for emergency warning and alerting. Comput. Environ. Urban Syst. **58**, 12–18 (2016). https://doi.org/10.1016/j.compenvurbsys.2016.03.003

13. Garcia, L., Escobar, C., Arango, E.: Sistemas de alerta temprana con enfoque participativo (2016). http://lunazul.ucaldas.edu.co/index.php/english-version/91-coleccion-articulos-espanol/235-sistemas-de-alerta-temprana-con-enfoque-participativo

14. Visheratin, A.A., Melnik, M., Nasonov, D., Butakov, N., Boukhanovsky, A.V.: Hybrid scheduling algorithm in early warning systems. Future Gen. Comput. Syst. **79**, 630–642 (2018). https://doi.org/10.1016/j.future.2017.04.002

Design, Implementation and Characterization of a Low-Cost Stair-Climbing and Obstacle Dodging Robot for Emergency Situations

Rafaela Villalpando-Hernandez, Cesar Vargas-Rosales[✉],
Rene Diaz-M., J. C. Armendariz-Nuñez, E. E. Castañeda-Hernandez,
and L. A. Bollinger-Rios

Tecnologico de Monterrey, Monterrey, Mexico
{rafaela.villalpando, cvargas, renejdm}@tec.mx,
julioarmendariz19@gmail.com,
erik.castaneda.sgh@gmail.com,
luis.bollinger.23@gmail.com

Abstract. Disaster response robots (DRRs) designed to explore dangerous areas in emergency situations (collapses, urban conflagration, earthquakes, floods) represent a design challenge. Therefore, there is the constant need of innovation in the field of navigation through harsh environment. In this paper, we present a design and implementation of a low cost, low power consumption and high terrain adaptability continuous track articulated robot. The proposed design is capable of climbing obstacles and dodging them by adjusting the inner angle of a middle joint. Also the current research, provides new ideas towards the innovation and ad-hoc use of low-cost robots for emergency scenarios.

1 Introduction

There are plenty of robot designs made to overcome different obstacles, go fast or adapt to different environments by their sturdiness, [1–3]. When designing these kinds of robots, it is important to choose a low-cost solution capable of navigating in dangerous areas, exploring different levels of a building, and overcoming objects in their paths. The requirements of disaster response robots (DRRs) are: dodging tall objects by veering, overcoming short obstacles to avoid jamming and having sufficient traction to climb stairs or medium height objects, as explained in [1]. In order to meet these requirements, the moving method chosen must be most suitable for those conditions.

Some basic methods used by the DRRs to move are wheels, arms, continuous tracks or a combination of these, as in [1, 4, 5]. Nevertheless, all these methods have some advantages and disadvantages, for example; wheels are often used due to their simplicity in installation and functionality. Arms are more difficult to implement, since they have more degrees of freedom, more force reactions to be computed, and they are relatively difficult to program, and slower, but they are more reliable in difficult terrains than wheels. Continuous tracks are usually implemented in unknown environments, where the terrain can present a large range of irregularities and therefore, a higher torque is required.

In this paper, we present a high terrain adaptable, easy to control, low cost and low power consumption robot capable of dodging obstacles that when unable to overcome.

A. Martínez et al. (Eds.): LACAR 2019, LNNS 112, pp. 236–247, 2020.
https://doi.org/10.1007/978-3-030-40309-6_23

Time and low economic resources were the main considerations for the final design. The design is divided into the choice of crank vs. wheels, the chassis and the slider crank mechanism. The contribution of our work is to provide a low complexity alternative (two degrees of freedom) for emergency support.

DRRs design utilizing a combination of wheels, tracks and arm or hybrid loco-motion such as arm-crawlers [1, 5], or a frame [4], have shown that it is necessary to combine various types of movements to overcome treacherous terrain as efficiently as possible. For example, wheeled robots excel in linear movement in even terrain, articulated robots are ideal for uneven terrains and legged robots are used to imitate human motion. In [1], authors present a DRR that by utilizing an arm-crawler mech-anism, the robot takes advantage of the maneuverability on an even terrain with a wheel, and by the inclusion of an arm, the robot gets the ability to climb obstacles by imitating a human limb. However, all the mechanism that conforms a robot has its mechanical and electrical limits, which can be improved by optimizing chassis, wheels and other mechanical component materials, trajectory control algorithms and an effi-cient power supply. It can be assumed that, by combining two or more of these improvements, a mid-point in mobility and maneuverability could be achieved and by doing so, it is possible to enhance the capabilities of the robot.

In Sect. 2, the proposed DRR design and implementation are discussed. Section 3 introduces the mechanical design of the robot's chassis. In Sect. 4, the main mecha-nism to overcome obstacles is discussed. In Sects. 5 and 6 the math of the robot and the movement algorithm are presented, respectively. Section 7 includes the conclusions.

2   Continuous Track vs. Wheels

In this section the design and implementation of a low cost, low complexity and low power consumption DRR is presented. When designing a DRR, the characteristics of the terrain and the purpose of the DRR, make necessary the combination of different methods to have efficient movement. For instance, stair climbing requires an arm-like mechanism to reach the top of the stairs. Also, it was necessary to implement the wheel or continuous track design alike. In the wheel design, every single one of the four wheels needs a motor and an arm, in order to climb the steps, making it 8 motors, where four of them need to be step motors. The continuous track design requires just four regular motors for the movement and one for the arm-like mechanism. In Table 1, the motor's specifications are described. Also, in Table 1 we provide specifications of the microcontroller (MPU 6050), ultrasonic proximity sensor (HC SR04) and the Bluetooth module (HC-05) used to complete the proposed DRR prototype.

Table 1. Drivers and devices specifications

	Voltage (V)	Current (mA)	RPMS
Slider-crank motor	6	400	30
Movement motor	6	250	12
MPU 6050	3.5	–	–
Ultrasonic HC SR04	5	15	–
Bluetooth HC-05	3.3	35	–

Even though, continuous tracks are slower and more complex to construct, the fact that they have more contact with the floor than wheels, represent the best low-cost solution. Small obstacles and irregular terrain can easily be overcome, making programming easier and faster. Another reason to not use wheels is the multiple steps stairs, where the length of the continuous tracks helps the upward movement, using the edge of every step as support. The mechanical design of the motion system implemented (not including the bike tire) is shown in Fig. 1.

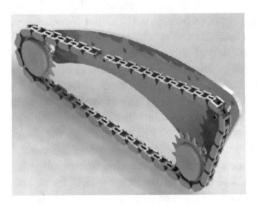

Fig. 1. Font continuous track design using ½-bike chain.

3   Chasis

As stated before, one of requirements is to design and implement a low cost DRR. Therefore, to reduce the motors' torque and power consumption, the robot's chassis was constructed using medium-density fiber board (MDF) wood, which is inexpensive and easy to cut. Continuous tracks are not meant to be fully tightened in order to adjust to the terrain's shape for more traction, leaving a curvy design to avoid blockades, see Fig. 2.

The scenario for the robot experimentation was defined by measuring regular stairs in a typical environment (e.g., university campus). Physical robot's dimensions can be seen in Fig. 3(a), where the top view of the robot is presented. Note that $l_1 = 300$ mm. In Fig. 3(b), we present the top view of the rear part of the DRR, in this figure we can observe that the separation between the front part of the chassis and the second track is defined as $l_2 = 100$ mm. Finally, in Fig. 3(c), the front view of the slider-crank mechanism can be observed, and distance from the first gear of the back track to the anchor of the slider crank mechanism is set as $l_3 = 240.96$ mm and $l_4 = 218.92 + 167.17 = 386.09$ mm for linear links. Due to the length of the continuous tracks and the DRR's width, the front of the robot must be lifted in order to veer left or right. To keep a low construction cost and a high sturdiness of the robot, cutting and reconstructing chain bike with a sixteen teeth sprocket joined with a self-machined coupler to the motor was the method used for the four continuous tracks. To increase traction in every terrain, we used rubber from a bike tire, joined to the bike chain.

Fig. 2. Curvy chassis to adjust to terrain.

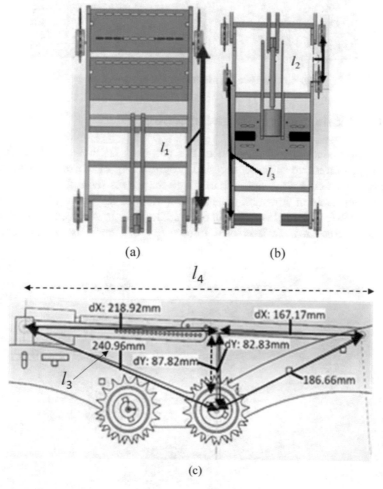

Fig. 3. (a) Top view of front part of DRR. $l_1 = 300$ mm. (b) Top view of rear part of DRR. $l_2 = 100$ mm. (c) Front view of slider-crank mechanism using worm screw.

4 Slider Cranck Mechanism

In order to lift the front part of the DRR without interacting with the terrain, we designed a slider-crank mechanism using a worm screw. This ensured high torque at slow speeds using a low power motor which characteristics are described in Table 1. In Fig. 4, the slider-crank mechanism implemented is shown.

Fig. 4. Slider-crank mechanism using worm screw.

5 Robot Mathematical Characterization

For the mathematical characterization of the proposed DRR, we took a simple but accurate approach to aid in the design and programming of the robot. Although, our robot began as an empiric design, the data and the results show that there's room for future improvements in structure, actuators and control drivers; it also validates our concept of a DRR with angle control in the middle joint. In sub-section A the forward kinematics is presented, as it is well known, by this analysis we can obtain the

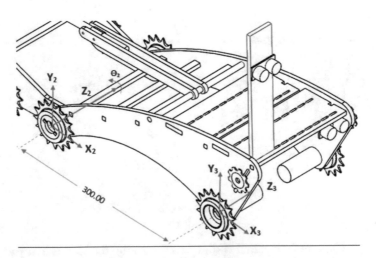

Fig. 5. Robot frames representing each independent movement (Dimensions given in millimeters).

localization of the endpoint of the robot in a determined instance of its path. In Fig. 5, we present the DRR mechanical design, where the reference axis can be observed, which will be used in the following kinematic analysis.

Table 2, contains the Denavit Hartenberg parameters, and the necessary variables for the forward and inverse kinematics, as well as the Jacob matrix. The information is obtained from Fig. 6, where the only unknown variable is the angle θ_2 generated by the worm screw, d_{i-1} is the distance from frame $i - 1$ to frame i in the z axis, α is the angle that needs to rotate around the x axis so frame i is parallel to frame $i - 1$, a is the distance moving on the x axis from frame $i - 1$ to frame i. And θ_{i-1} is the angle that spins around z_{i-1}, so x_{i-1} and x_i become parallel, [6].

Table 2. Denavit-Hartenberg parameters of frames 2 and 3.

I	α_{i-1}	a_{i-1}	d_i	θ_{i-1}
2	0	0	0	θ_2
3	0	300	0	0

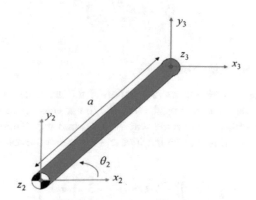

Fig. 6. Geometric analysis of robot frame 2 and 3.

Since the axis that joins the rear and back part of the DRR is parallel to every continuous track axis, there's no angle α. Constant a is the distance between both z-axis, which are pointing outside the page (represented by a point), of frame 2 and 3. θ_2 is the angle of rotation around z-axis of frame 2, [6].

5.1 Forward Kinematics

Forward kinematics analysis provides information about location of the end frame, when variables of the robot changes. In this case, there's only one variable, which is θ_2. In Eq. (1) we present the Denavit Hartenberg transformation matrix A_1 which is used to calculate z coordinate upon θ_2, resulting in different x and y coordinate values. That represents a 1 degree of freedom robot, if we only focus on the step climbing capabilities, excluding others like veering or frontal movement.

$$A_1 = \begin{bmatrix} cos\theta_2 & -sin\theta_2 & 0 & 300cos\theta_2 \\ sin\theta_2 & cos\theta_2 & 0 & 300sin\theta_2 \\ 0 & 0 & 1 & 0 \\ 0 & 0 & 0 & 1 \end{bmatrix} \tag{1}$$

Recall that in the Denavit Hartenberg, method the homogenous transformation matrix A_1, that represent the translation and rotation transformation matrix corresponding to the z axis and the translation and rotation transformation matrix corresponding to the x axis.

5.2 Inverse Kinematics

Using the inverse kinematics analysis, we can calculate the exact angle θ_2 of the robot's middle joint, which gives possible terrain insight. If the terrain is similar or known for the robot, then, it is possible to apply the same terrain analysis and this analysis can be used for a quick adjustment in the control algorithm.

$$\theta_2 = tan^{-1}\left(\frac{dy}{dx}\right) \tag{2}$$

5.3 Jacobian for Velocity Analysis

Finally, the Jacobian analysis allows to calculate if the movement of the robot is being made correctly. The Jacobian matrix gives us the linear velocity in the axes ($\dot{x}$, $\dot{y}$ and $\dot{z}$) and the angular velocities in the same way (ω_x, ω_y and ω_z). These angular velocities can be used for further correction in the motor drive when the robot is going downhill or stepping down.

$$\begin{bmatrix} \dot{x} \\ \dot{y} \\ \dot{z} \\ \omega_x \\ \omega_y \\ \omega_z \end{bmatrix} = \begin{bmatrix} -300sin\theta_2 & -300sin\theta_2 \\ 300cos\theta_2 & 300cos\theta_2 \\ 0 & 0 \\ 0 & 0 \\ 0 & 0 \\ 1 & 1 \end{bmatrix} [\theta_2] \tag{3}$$

6 Motion

The robot is operated via Bluetooth, using the HC-06 device (see Table 1 for specifications), that allows connection between the microcontroller ATmega328P and a smartphone. When turned on, the robot begins to drive the four continuous tracks, which, as mentioned before, can go up steps and turn left or right. In Fig. 7, we provide a block diagram describing the robot's movement logic.

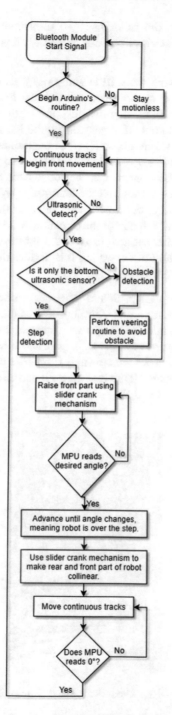

Fig. 7. Algorithm for obstacle dodging.

Given these conditions, the robot has two ultrasonic sensors that are used to measure the proximity of obstacles like walls or steps. If it is in front of a wall, there are three possibilities:

- If the shortest way to avoid the wall is going right, using the slider-crank mechanism, it rises the front part of the robot, this in order to allow easier veering due to the DRR's narrowness. Then, it turns only the right motors on and after some seconds, it turns those motors off, then turns on the left motors the same amount of time that the right motors were on, so it can be positioned at the same direction as it was at the beginning, but at a considerable distance to the right. Later, the slider-crank mechanism puts the front part of the robot back down.
- If the robot is in front of a wall and the shortest way to avoid it is going left, it follows the same procedure, but now it will move to the left.
- If the robot can't detect which one of the directions (left or right) is the shortest, the robot will follow the same instructions as if the shortest way is to the right, then it evaluates if it is still in front of the wall, if it is, it does the same actions again, if it's not, it continues to move forward.

When the robot detects a step, this is, if the top ultrasonic sensor reads a few centimeters more than the bottom one (Fig. 8a), it uses the slider-crank mechanism to rise up to a fixed optimal angle (this angle was measured using an MPU sensor), see Fig. 8b. Then it moves forward until the front part is on the edge of the step, then the worm screw is activated to make the rear and front part of the DRR collinear again. Finally, the robot turns on again all the motors, so it can climb the step or steps. This is shown in Figs. 9a, 9b, 9c, 9d and 9e.

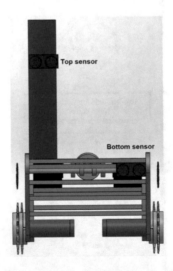

Fig. 8a. Front view of the DRR, where the top and bottom ultrasonic sensor can be seen.

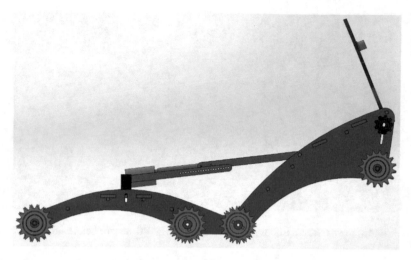

Fig. 8b. Position taken to allow easier veering (rear part of the DRR stores the battery and slider motor, making it heavier).

Fig. 9a. Lower ultrasonic sensor detects a step, while upper one detects nothing.

Fig. 9b. Robot starts rising its front part by sending a signal to the screw system motor.

Fig. 9c. Screw motor stops. Robot advances until the MPU sensor returns the desired angle value.

Fig. 9d. The screw motor makes the rear and front part collinear again. This in order to allow a safe step climbing.

Fig. 9e. Finally, motion motors turn the continuous tracks on, and robot advances until MPU sensor reads zero degrees.

Every robot's movement, is sequential, never do more than one movement at a time, this in order to maintain easy control and low current consumption. A table showing the performance in terms of power consumption for each independent movement of the DRR, excluding friction losses, can be seen in Table 3.

Table 3. Power consumption.

Movement	Power (W)
Front part lifting	2.4
Veering	3.0
Forward drive	6.0

7 Conclusions

In this paper, we present a low complexity, low power consumption and low cost DRR robot, these characteristics make it easy to debug, control, and modify. Throughout experimentation the functionality of the continuous tracks was demonstrated. It was also shown that the DRR has low power consumption, However, power consumption may be improved adding PID controllers to every motor, smoothing movements and reducing current peaks. This modification can even make several movements at a time possible keeping low power consumption. Therefore, this paper presents a proof of concept of a low cost DRR design and characterization, and for future modifications, the sturdiness and reliability would be the main consideration.

References

1. Chen, K., et al.: Compound locomotion control system combining crawling and walking for multi-crawler multi-arm robot to adapt unstructured and unknown terrain. Robomech **5**, 17 (2018)
2. Coetzee, S., Bosscha, P., Swart, H., Oosthuizen, D.: Design of an industrial all-terrain robot platform. In: RobMech, Pretoria, November 2012
3. Hata, M.A.A.B.M., Baharin, I.: The study of a customisable all terrain mobile robot (ROBUST). In: 2014 11th International Conference on Ubiquitous Robots and Ambient Intelligence (URAI), Kuala Lumpur, pp. 232–237, November 2014
4. Mishra, P., Adithya, G., Kishore, J.K.: Design and characterization of two-frame all-terrain robot for navigation in unstructured and treacherous environments. In: 2012 Annual IEEE India Conference (INDICON), Kochi, pp. 155–160, December 2012
5. Ben-Tzvi, P.: Experimental validation of a hybrid mobile robot mechanism with interchangeable locomotion and manipulation. In: 2009 IEEE/RSJ International Conference on Intelligent Robots and October Systems, St. Louis, MO, pp. 420–421 (2009)
6. Craig, J.: Robotics, 3rd edn. Pearson, London (2006)

Dynamics and Preview Control
of a Robotics Bicycle

Diego A. Bravo M.$^{(\boxtimes)}$, Carlos F. Rengifo R., and Wilber Acuña B.

Universidad del Cauca, Calle 5 Nro 4-70, Popayán, Colombia
{dibravo,caferen,wacuna}@unicauca.edu.co

Abstract. This paper presents the dynamic model of the Arduino Engineering Kit robotics bicycle. We design and implement a closed loop control strategy (preview control), that minimizes the tracking error and rejects disturbances generated by uncertainties and unmodeled dynamics. The *preview control* is a servo controller that permits the tracking of desired trajectories in the handlebar, maintaining the stability of the bicycle. Finally, we tested the proposed control strategy on a real robotic bicycle prototype.

1 Introduction

Bicycles are open loop unstable systems requiring that the human driver controls the steering and balances his body to stabilize the system. Therefore, it is of scientific and engineering interest to study bicycle stabilization systems.

From a pedagogical perspective, bicycles can be used to illustrate a wide variety of concepts in control engineering [4]. Bicycles have been used as cases of study in courses of dynamic systems and control engineering, as it is the case of the universities of Lund, California and Cornell [17].

A two-wheeled bicycle shares the same characteristics as an inverted pendulum, [4]. It is non-linear, unstable, multivariate and strongly coupled [2]. These features have increased the interest of many researchers in to provide teaching and research tools that allow to consolidate the knowledge related to different control strategies, signal processing and the obtaining of kinematic and dynamic models. Currently they are counted with some models for control purposes such as Åström [4] and Limebeer [15].

This type of vehicle is affected by external disturbances, unmodeled dynamics, errors in the estimation of the parameters and the noise in the measurements obtained by the sensors, among others [18].

Most of the literature about control of bicycles is devoted to the application of both linear and non-linear control techniques to these systems. In general, these techniques require of a mathematical model [8] that can be obtained by using Lagrange equations [7], or the Newton-Euler algorithm [16], or the Newton's laws [12]. For the linear control strategies, the model needs to be linearized, contrary to the non-linear control, where the mathematical model can directly used.

© The Editor(s) (if applicable) and The Author(s), under exclusive license to
Springer Nature Switzerland AG 2020
A. Martínez et al. (Eds.): LACAR 2019, LNNS 112, pp. 248–257, 2020.
https://doi.org/10.1007/978-3-030-40309-6_24

The literature presents control techniques for the stabilization of a bicycle such as: PID Control [6], state feedback [3], nonlinear control [20], model-based Active Disturbance Rejection [19], model predictive control [1], among others.

The main objective of this work is to implement a control strategy that permits the tracking of desired steer trajectories, besides of to stabilize the inclination of a robotic bicycle that moves freely at a constant speed. To achieve this goal, we obtain the mathematical model of the *Arduino Engineering Kit* prototype by means of Lagrange equation. The control proposed, *preview control* utilizes future information of the references trajectories in off-line mode [9,11,13], and minimizes the tracking error and rejects disturbances generated by uncertainties and unmodeled dynamics.

The rest of the paper is structured as follows: Sect. 2 describes the hardware of the system and the kinematic model of bicycle. Section 3 shows the dynamics model of a robotic bicycle, Sect. 4 presents the preview controller implemented on the dynamic system, followed by the results in Sect. 5 and concluding paragraph of the paper in Sect. 6.

2   Robotic Bicycle

The robotic bicycle (see Fig. 1) from *Arduino Engineering Kit* is composed by a micro servo motor in the steer column, one gear motor with encoder to move the back wheel, a DC motor to make the disc spin, one ultrasonic sensor, and three electronic boards. The main board contains the controller and permits Wi-Fi connectivity, the second board is an *Inertial Measurement Unit* (IMU) and the third board is a power interface for the motors. Although each body part of the prototype can be disassembled and its mass measured, the center of mass of each body part and thus the entire robot is harder to predict.

2.1   Kinematic Model

In order to describe the kinematics of the system, the reference systems of Fig. 2 are defined. The inertial frame is $<x_g, y_g, z_g>$, while $<x, y, z>$ is the frame of contact point of the wheel. ψ is the angle of rotation around the axis y_g needed to match the axis x_g with the axis x, θ is the angle of rotation around the axis x_g needed to match the y_g axis with the y axis. From the definition of reference systems, we have:

$$\dot{x} = v \cos \psi$$
$$\dot{z} = v \sin \psi \qquad\qquad (1)$$
$$\dot{\psi} = \frac{v}{\rho} = \frac{v \tan \phi}{b}$$

$\rho = \frac{b}{\tan \phi}$ es is the *radius of curvature* of the path the bicycle follows when the angle that the direction makes with respect to the axis x is ϕ.

Fig. 1. Robotic bicycle from *Arduino Engineering Kit*.

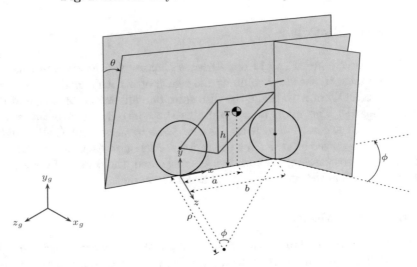

Fig. 2. Geometric model.

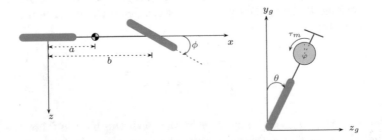

Fig. 3. Top and rear views.

3   Dynamic Model of Robotic Bicycle

The reference systems for the dynamic model of the bicycle are shown in Fig. 4. The inertial frame is $<x_g, y_g, z_g>$, the generalized coordinates used to describe the motion are: x for the translation, θ denote the inclination of the bicycle, φ is the rotation angle of gyro wheel and ϕ is the rotation angle of helm.

By using the Lagrange formalism the following mathematical model is obtained:

$$\mathbf{I}(q)\,\ddot{q} + \mathbf{C}(q,\dot{q})\,\dot{q} + \mathbf{G}(q) = \mathbf{\Gamma} \tag{2}$$

In the previous model $q \in I\!R^5$ is the vector of generalized coordinates. This vector contains two types of coordinates: (i) $(^g x_0, {}^g y_0, {}^g z_0)$ denoting the Cartesian position of the reference frame $<x_0, y_0, z_0>$ in the reference frame $<x_g, y_g, z_g>$, (ii) θ, φ, ϕ are the other coordinates to define the movement of the bike.

$$q = \begin{bmatrix} {}^g x_0 & {}^g z_0 & \theta & \varphi & \phi \end{bmatrix}^{\mathrm{T}}$$
$$\dot{q} = \begin{bmatrix} {}^g \dot{x}_0 & {}^g \dot{z}_0 & \dot{\theta} & \dot{\varphi} & \dot{\phi} \end{bmatrix}^{\mathrm{T}} \tag{3}$$

$\dot{q} \in I\!R^5$ is the vector of velocities generalized and $\ddot{q} \in I\!R^5$ is the vector of acceleration generalized. $\mathbf{I} \in I\!R^{5 \times 5}$ is the robot's inertia matrix. $\mathbf{C} \in I\!R^5$ is the vector of centrifugal and Coriolis forces, $\mathbf{G} \in I\!R^5$ is the vector of gravitational forces. $\mathbf{\Gamma} \in I\!R^5$ is the vector of torques applied to the joints of the robot.

$$\mathbf{I} = \begin{bmatrix} m_0 + m_1 + m_2 & 0 & 0 & 0 & 0 \\ 0 & m_0 + m_1 + m_2 & -\sin\theta\,(h\,m_0 + m_1\,y_1 + m_2\,y_2) & 0 & 0 \\ 0 & -\sin\theta\,(h\,m_0 + m_1\,y_1 + m_2\,y_2) & m_0\,h^2 + m_1\,y_1{}^2 + m_2\,y_2{}^2 + J_0 + J_1 & J_1 & 0 \\ 0 & 0 & J_1 & J_1 & 0 \\ 0 & 0 & 0 & 0 & J_2 \end{bmatrix}$$

$$\mathbf{C} = \begin{bmatrix} 0 \\ -\dot{\theta}^2\,\cos\theta\,(h\,m_0 + m_1\,y_1 + m_2\,y_2) \\ 0 \\ 0 \\ 0 \end{bmatrix} \qquad \mathbf{G} = \begin{bmatrix} 0 \\ g\,(m_0 + m_1 + m_2) \\ -g\,\sin\theta\,(h\,m_0 + m_1\,y_1 + m_2\,y_2) \\ 0 \\ 0 \end{bmatrix}$$

$$\mathbf{\Gamma} = \begin{bmatrix} \tau_g/r & 0 & 0 & \tau_m & \tau_d \end{bmatrix}^{\mathrm{T}}$$

3.1   CAD Model

The parameters of the dynamic model were derived from the CAD models of the robotic bicycle. These CAD specifications include the inertia moments and center of mass of each body. The CAD model does not include the mass of elements such as electronic boards, wiring, and hollow spaces in plastic body. The obtained model does not take into consideration the mechanical backlash in the gears, nor the non-linearities in the electrical motor [5]. These phenomena mean that idealized position controlled motors in a simulation can act differently from those in the real world unless the controller is designed to compensate these sources of error. The parameters of the model described by Eq. (2) are shown in the Table 1.

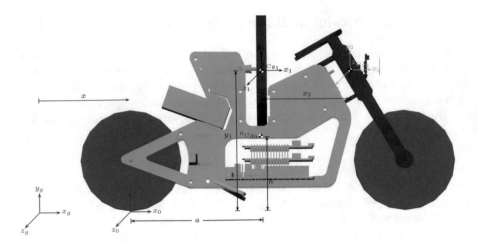

Fig. 4. Reference systems.

Table 1. Dynamic parameters

Parameter	Symbol	Value
Bike Mass	m_0	$0.3227\,\text{kg}$
Wheel Mass	m_1	$0.0764\,\text{kg}$
Helm Mass	m_2	$0.0773\,\text{kg}$
Center of Mass	(a, h)	$(0.11,\ 0.07)\,\text{m}$
Bike Inertia moment	J_0	$0.0049\,\text{kg} \cdot \text{m}^2$
Wheel Inertia moment	J_1	$0.00062\,\text{kg} \cdot \text{m}^2$
Helm Inertia moment	J_2	$0.00029\,\text{kg} \cdot \text{m}^2$
Wheel radius	r	$0.05\,\text{m}$
Distance	y_1	$0.13\,\text{m}$
Distance	y_2	$0.13\,\text{m}$
Gravity	g	$9.81\,\text{m} \cdot \text{s}^{-2}$
Motor Torque	τ_g	$0.217\,\text{N} \cdot \text{m}$
Motor Torque	τ_m	$0.014\,\text{N} \cdot \text{m}$
Motor Torque	τ_d	$0.225\,\text{N} \cdot \text{m}$

4   Preview Control System

A controller which utilizes future information is called *preview control* [9,11,13].

The Lagrange model, Eq. (2) has been linearized using the Jacobian Matrix, [10] and discretized with a sample time h. With the state vector $\mathbf{x}(k) = [x,\, z,\, \theta,\, \varphi,\, \phi]^T$ the dynamics equation are:

$$\mathbf{x}(k+1) = A\,\mathbf{x}(k) + B\,\mathbf{u}(k)$$
$$\mathbf{y}(k) = C\,\mathbf{x}(k)$$

$$(4)$$

Where, $A \in I\!R^{10 \times 10}$, $B \in I\!R^{10 \times 3}$, $C \in I\!R^{2 \times 10}$ and $\mathbf{u}(k) = [T_g/r, T_m, T_d]^T$. To let the system output $\mathbf{y}(k)$ follow the desired trajectory y_k^d in the plane $<x, z>$ as closely as possible. We designed a servo controller for the system Eq. (4) to minimize the following performance:

$$J(k) = \sum_{j=k}^{\infty} \left[Q \left(y_j^d - y_j \right)^2 + R u_j^2 \right] \tag{5}$$

The parameters of matrices Q and R are designed to penalize deviations on the state variables and control signals, respectively; since all variables are considered independent, these matrices become diagonal weights. Where Q, R are positive weights matrices. This is called a *tracking control problem*. The performance index J can be minimized by the following input which uses the future references up to N steps, [14].

$$u(k) = -G_i \sum_{i=0}^{k} e(i) - G_x \mathbf{x}(k) - \sum_{j=1}^{N_l} G_p(j) \, y^d(k + j) \tag{6}$$

$$e(i) = y(i) - y^d(i)$$

where

$$\mathbf{G} = \begin{bmatrix} G_i \\ G_x \end{bmatrix} = \left(R + B^T \mathbf{P} B \right)^{-1} B^T \mathbf{P} A$$

$$G_p(j) = \left(R + B^T \mathbf{P} B \right)^{-1} B^T \left(A - B \, \mathbf{G} \right)^{T*(j-1)} C^T Q \tag{7}$$

The matrix $\mathbf{P}$ is a solution of *Ricatti Equation*.

$$\mathbf{P} = A^T \mathbf{P} A + C^T Q C - A^T \mathbf{P} B \left(R + B^T \mathbf{P} B \right)^{-1} B^T \mathbf{P} A \tag{8}$$

The *preview controller* consist of three terms: the integral action on the tracking error $e(i)$, the state feedback and a feed-forward of a inner product between the future target reference up to N steps and the weights $G_p(j)$. A block diagram of the preview control is illustrated in Fig. 5. The future references y_{k+j}^d are stored in a **FIFO** (**First-In-First-Out**) buffer, and its output value is regarded as the current reference. The preview controller calculates the control input using the reference on the FIFO buffer and the state vector.

In this control approach, higher gain was assigned to the lean angle and steer angle, in order to guarantee the stability of bicycle and to make the tracking of the trajectory, respectively. Using the linear dynamic model Eq. (4) a controller for robotic bicycle was achieved by adjusting the matrices Q and R, which are related to the states and the inputs of system respectively. For the controller implementation, $Q \in I\!R^{2 \times 2}$ a positive semi-definite matrix of and $R \in I\!R^{2 \times 2}$, $N = 30$.

$$Q = \begin{bmatrix} 100 & 0 \\ 0 & 100 \end{bmatrix} \qquad R = \begin{bmatrix} 0.0001 & 0 \\ 0 & 0.0001 \end{bmatrix} \tag{9}$$

Using the linear dynamic model Eq. (2) the position controller for robotic bicycle was achieved by adjusting the matrices Q and R, which are related to the states and the inputs of system respectively.

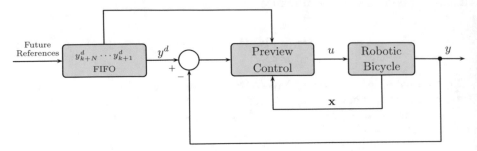

Fig. 5. Block diagram of *preview control.*

5   Experimental Results

The results obtained from the implementation of the *preview controller* are shown in Figs. 6 and 7. Figure 6 shows the tracking of reference signal for the steer angle ϕ and the control effort τ_d associated. The preview control permits

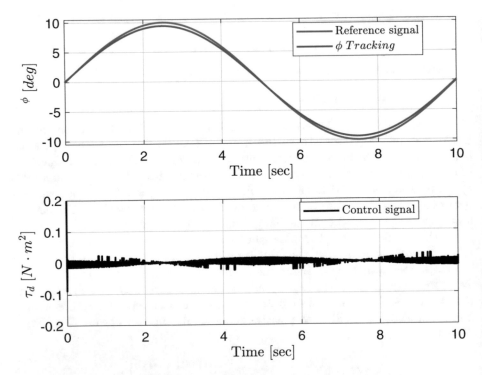

Fig. 6. Time response of the controller (steer angle ϕ) and control effort τ_d.

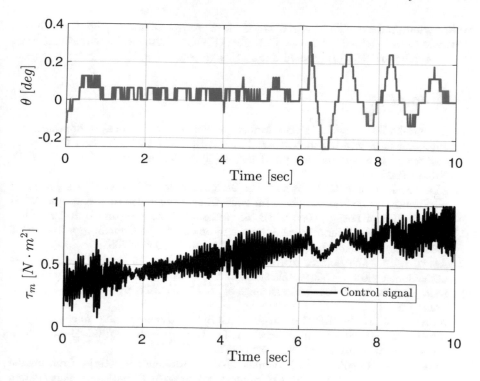

Fig. 7. Time response of the controller (lean angle θ) and control effort τ_m.

to use information about the future location of the bike for steering, this can be regarded as using future information for smooth autonomous driving.

The control input τ_m specifies the angular acceleration $\ddot{\varphi}$ of the bicycle about the ground-wheel axis is in the opposite direction of the lean angle θ. Specifically, when the bicycle leans one way, it will accelerate in the opposite direction (see, Fig. 3). When the bicycle moves to the other side of the equilibrium point, the angular acceleration $\ddot{\varphi}$ will then switch directions to move the bicycle back toward equilibrium $\theta_d = 0°$, like in the Fig. 7. This is a stable equilibrium, which means when the system is perturbed from the equilibrium position, it will restore itself back toward equilibrium when the bicycle moves at constant speed.

6 Conclusion

In this paper the dynamical model of a robotic bicycle was obtained by using the Lagrange formalism. The model was linearized by means of Jacobian matrix. A *preview control* was implemented using future information of desired trajectory for the steer and maintaining the balance of the bike. In future works, we will try to implement a model predictive control for the tracking of trajectories.

Acknowledgements. The authors would like to recognize and express their sincere gratitude to Universidad del Cauca, Colombia (UNICAUCA) for the financial support (Grant number 5116/2019) granted during this project.

References

1. Ai-Buraiki, O., Thabit, M.B.: Model predictive control design approach for autonomous bicycle kinematics stabilization. In: 22nd Mediterranean Conference on Control and Automation, pp. 380–383 (2014). https://doi.org/10.1109/MED.2014.6961401
2. Akesson, J., Blomdell, A., Braun, R.: Design and control of YAIP - an inverted pendulum on two wheels robot. In: 2006 IEEE Conference on Computer Aided Control System Design, 2006 IEEE International Conference on Control Applications, 2006 IEEE International Symposium on Intelligent Control, pp. 2178–2183 (2006). https://doi.org/10.1109/CACSD-CCA-ISIC.2006.4776978
3. Assad, M.M., Meggiolaro, M.A., Neto, M.S.: Analysis of control strategies for autonomous motorcycles stabilization and trajectories tracking. In: Fleury, A., Rade, D., Kurka, P. (eds.) Proceedings of DINAME 2017, pp. 401–416. Springer, Cham (2019)
4. Åström, K.J., Klein, R.E., Lennartsson, A.: Bicycle dynamics and control: adapted bicycles for education and research. IEEE Control Syst. **25**(4), 26–47 (2005). https://doi.org/10.1109/MCS.2005.1499389
5. Bravo, D.A., Rengifo, C.F.: Dynamics filter for walking trajectories from human motion capture. In: 2016 IEEE Colombian Conference on Robotics and Automation (CCRA), pp. 1–6 (2016)
6. Bravo, M.D.A., Rengifo, R.C.F., Díaz, O.J.F.: Control of a robotic bicycle. In: 2018 IEEE 2nd Colombian Conference on Robotics and Automation (CCRA), pp. 1–5 (2018). https://doi.org/10.1109/CCRA.2018.8588132
7. Consolini, L., Maggiore, M.: Control of a bicycle using virtual holonomic constraints. Automatica **49**(9), 2831–2839 (2013). https://doi.org/10.1016/j.automatica.2013.05.021
8. Edelmann, J., Haudum, M., Plochl, M.: Bicycle rider control modelling for path tracking. IFAC-PapersOnLine **48**(1), 55 – 60 (2015). https://doi.org/10.1016/j.ifacol.2015.05.070. 8th Vienna International Conference on Mathematical Modelling
9. Helin, W., Chengju, L., Qijun, C.: Omnidirectional walking based on preview control for biped robots. In: 2016 IEEE International Conference on Robotics and Biomimetics (ROBIO), pp. 856–861 (2016). https://doi.org/10.1109/ROBIO.2016.7866431
10. Horla, D., Owczarkowski, A.: Robust LQR with actuator failure control strategies for 4DoF model of unmanned bicycle robot stabilised by inertial wheel. In: 2015 International Conference on Industrial Engineering and Systems Management (IESM), pp. 998–1003 (2015). https://doi.org/10.1109/IESM.2015.7380276
11. Huaman-Loayza, A.S.: Accurate trajectory tracking for a 3D-plotter using optimal preview control and unknown input observer. In: 2018 IEEE XXV International Conference on Electronics, Electrical Engineering and Computing (INTERCON), pp. 1–4 (2018). https://doi.org/10.1109/INTERCON.2018.8526428

12. Hung, N.B., Sung, J., Lim, O.: A simulation and experimental study of operating performance of an electric bicycle integrated with a semi-automatic transmission. Appl. Energy **221**, 319–333 (2018). https://doi.org/10.1016/j.apenergy.2018.03.195
13. Jiang, S., Zhen, Z., Jiang, J., Yang, Z.: A preview control scheme for carrier-based aircraft automatic landing. In: 2018 37th Chinese Control Conference (CCC), pp. 9815–9819 (2018). https://doi.org/10.23919/ChiCC.2018.8483965
14. Kajita, S., Hirukawa, H., Harada, K., Yokoi, K.: Introduction to Humanoid Robotics, vol. 1. Springer, Heidelberg (2014). ISSN 1610-7438. Original Japanese edition published by Ohmsha Ltd., Tokyo (2005)
15. Limebeer, D.J.N., Sharp, R.S.: Bicycles, motorcycles, and models. IEEE Control Syst. **26**(5), 34–61 (2006). https://doi.org/10.1109/MCS.2006.1700044
16. Mauny, J., Porez, M., Boyer, F.: Symbolic dynamic modelling of locomotion systems with persistent contacts - application to the 3D bicycle. IFAC-PapersOnLine **50**(1), 7598 – 7605 (2017). https://doi.org/10.1016/j.ifacol.2017.08.1007. 20th IFAC World Congress
17. Papadopoulos, J.: Bicycle steering dynamics and self-stability: a summary report on work in progress. Technical report, Cornell University (1987)
18. Raffo, G.V., Ortega, M.G., Madero, V., Rubio, F.R.: Two-wheeled self-balanced pendulum workspace improvement via underactuated robust nonlinear control. Control Eng. Pract. **44**, 231–242 (2015). https://doi.org/10.1016/j.conengprac.2015.07.009
19. Tamayo-León, S., Pulido-Guerrero, S., Coral-Enriquez, H.: Self-stabilization of a riderless bicycle with a control moment gyroscope via model-based active disturbance rejection control. In: 2017 IEEE 3rd Colombian Conference on Automatic Control (CCAC), pp. 1–6 (2017). https://doi.org/10.1109/CCAC.2017.8276434
20. Yi, J., Zhang, Y., Song, D.: Autonomous motorcycles for agile maneuvers, part II: control systems design. In: Proceedings of the 48h IEEE Conference on Decision and Control (CDC) held jointly with 2009 28th Chinese Control Conference, pp. 4619–4624 (2009). https://doi.org/10.1109/CDC.2009.5399525

Design of a Low-Cost Ball and Plate Prototype for Control Education

Cristian Ramirez[✉], Paula Hurtado, Cristian Sanabria, and Kevin Ramirez

UPTC, Tunja, Colombia
{cristian.ramirez02,paulaandrea.hurtado,cristian.sanabria01,
kevin.ramirez}@uptc.edu.co

Abstract. This paper presents design and control of a low-cost prototype Ball and Plate as learning resource tool, using second-hand components and Perspex pieces. A digital controller allows defining a specific position of a Ball and Plate. The position is sensed by using a flat which has a resistive sensor. A PID control by poles allocated was used. The Ball and Plate platform allows the operator to interact through an application and observe the reaction variable behavior by using an android device.

1 Introduction

Some problems related with control require contributions from different disciplines, because these system are covered in a wide range of applications like design and manufacture of industrial devices or equipment, areas of biotechnology, renewable energy sources, etc. [2]. The ball and plate system belongs to a class of low-powered systems and is used to study the control algorithms, it is an unstable open loop system as the position of the ball is without limits when the plate tilts around of any of its axes. The coupling between the two axes is another limitation which adds to the complexity of the system.

Because the movement of the sphere is given in two dimensions, to address this problem and there is a solution with a conventional control that is applied are PID controls and is controlled for each variable, having as disadvantage that the two variables are not related We cannot offer an optimal response or a fine movement of the sphere on the plate, to improve this type of response, plant variations of PID and PI controllers to relate the error of each of the position variables and improve the response [1]. However, there is no solution, but also a development, a system of multiple variables, such as RVE, LQR, LQG, which relate all the variables within the system and have a better response in this case. It comes in the fine movement of the sphere above and plate [5], for the implementation of these systems it is necessary to obtain the closest possible plant model [6]. The Android platform and its user applications now dominate the field of mobile computing, including smartphones, tablets and other consumer electronics [3], which in addition to the control field allows greater versatility among resources in the development of monitoring through the interface, this being a current trend as potential development in home automation and

A. Martínez et al. (Eds.): LACAR 2019, LNNS 112, pp. 258–265, 2020.
https://doi.org/10.1007/978-3-030-40309-6_25

security in the industry, remote monitoring, etc. So applying an Android platform (whose concepts will be discussed in the course of digital electronics) to
the ball and plate structure strengthens the character of its development along
with the work of the control systems.

Based on the PBL technique [4], the proposed 2DOF test bench is developed
based on a square plexiglass plate that supports a 7-inch resistive touch sensor.
It is fixed on a rotating joint adapted from an axial directional ball joint that
performs movements on the plate by means of the position loop of two Futuba
S3OO3 servo motors, which includes the plate in response to the dynamic movement in order to stabilize the ball and not allow it to come out of the control
area, obtaining a platform composed of reused components and low-cost components compared to the costs of the platforms in the market, which allows
control techniques to be tested on a versatile platform and available to learners
of control.

2   Modelling

The modelling of the Ball and Plate system is approached through an analysis strategy that allows validating its model, composed of two parts: the first
includes the interaction of the axes (x and y) as a single dynamic system type
MIMO and the second decouples the system in two subsystems ball and beam
type SISO, then, the second part is presented where an analysis of the energies
of the system is performed using the Lagrange method (Fig. 1).

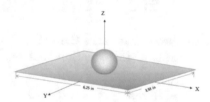

Fig. 1. Ball and Plate system has a rotation limited to 30° on the x and y axes. The
lengths of the plate in the axes correspond to Lx = 3.93 in and Ly = 6.25 in.

Parameters System

Ec:	Kinetic Energy of the system
Ep:	Potential energy of the system
I:	Rotational Inertia
w:	Angular speed
V:	Linear speed
m:	Traslational inertia (0.125 gr)
R:	Ball radius (0,015m)
d:	Servo motor arm radius (0,0165m)

The mathematical tool that allows the analysis of system energies is the Lagrange equation that relates the potential energy and the kinetic energy of the system, in the case of the ball:

$$L(t) = E_c(t) - E_p(t) \tag{1}$$

Kinetic energy is the sum of rotational energy and transnational energy, as described in Eqs. 2 and 3,

$$E_c(t) = E_{rot} + E_{tras}; \tag{2}$$

$$E_c(t) = \frac{1}{2}J\frac{\dot{r}^2}{R^2} + \frac{1}{2}m\dot{r}^2; \tag{3}$$

The potential energy of the ball is the gravitational potential energy that is determined as:

$$E_p(t) = -mg * sen(\alpha) \tag{4}$$

Substituting into the Lagrange equation,

$$L(t) = \frac{1}{2}J\dot{r}^2 + \frac{1}{2}m\dot{r}^2 + mg\sin(\alpha) \tag{5}$$

As we already know the Lagrange equation of the system we can obtain the equations of movement by means of the direct substitution of the expression for the Lagrange function in the Euler-Lagrange equation, which is shown below

$$\frac{d}{dt}\left(\frac{\delta L}{\delta t}\right) - \frac{\delta L}{\delta L} = F \tag{6}$$

Since there are no external forces that cause the ball to move, we have F = 0 in this way evaluating the derivatives the equation that represents the movement of the system is:

$$\left(\frac{J}{R^2} + m\right)\ddot{r} - mg\sin(\alpha) = 0 \tag{7}$$

The next step is to linearize the model. For this, a point of equilibrium is defined, in which it is possible to linearize the expression as follows:

$$sen(\alpha) = \alpha \tag{8}$$

For small variations of the α angle,

$$\frac{J}{R^2 + m}\ddot{r} - mg\sin(\alpha) = 0 \tag{9}$$

The Eq. 9, models the behavior of the ball on the plate, but the variable that interests us to be able to control the position of the ball is the angle θ that is why it is necessary to change the variable from α to θ. by the physical construction of the plant the arc length that the dish moves must be proportional to the arc length that the servomotor arm rises, so that:

$$L\alpha = d\theta \tag{10}$$

Substituting,

$$\left(\frac{J}{R^2} + m\right)\ddot{r} = -mg\frac{d\theta}{L} \tag{11}$$

$$\frac{R(s)}{\theta(s)} = -\frac{mgd}{L(\frac{J}{R^2} + m)s^2} \tag{12}$$

This is the transfer function the decoupled system, where for each axis the value of the constant will be different.

$$G(s) = \frac{km}{s^2} \tag{13}$$

3 Prototype Design

The design of the physical structure of the didactic platform ball and plate needs a base, support parts for servomotors, support for the plate on which the sphere will move, arms that transmitte the movement to the servomotor to the plate and an axis that will have a joint of two degrees of freedom that allows the movement of inclination and balance of the plate.

It is decided to manufacture the support pieces in acrylic using laser cutting, for this purpose, SolidWorks software is used where the design of: the base, taking into account the total dimensions of the structure. The support for the plate according to the dimensions of the sensor. The supports for the servomotors, taking into account the location of the arms with the center of each of the axes, the coupling with the base of the platform and the dimensions of the servomotors.

Figure 2, you can see the drawings of the designs made for the servomotor supports, the shape of the base is square and the support of the plate is rectangular with the same dimensions as the sensor.

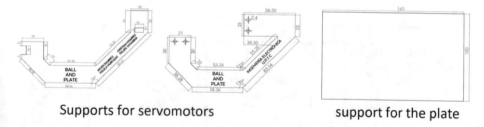

Supports for servomotors   support for the plate

Fig. 2. Plans of the fighters designed as support

Regarding the instrumentation, futaba s3003 servomotors are used, which provide enough torque for the task they perform, taking into account that the ball to be used is metal, using a 4-wire resistive tactile sensor that allows obtaining the coordinates of the ball on the plate thanks to the pressure exerted and as a controller the ATmega 328 is used in the Arduino development card, which facilitates the implementation of the controller, the reading of data and, most importantly, provides a stable control signal.

Fig. 3. Prototype ball and plate implemented and put into operation, the structure is portable which allows to be exposed in academic places.

4 PID Controller by Allocated Poles

It has been determined to implement a discrete PID controller to implement it in the microcontroller. The method used for the design of the controller is poles allocated by a desired model, so that a desired behavior of the system is proposed and it is tried to reach it by means of the controller.

Figure 4, shows a block diagram of the closed-loop control system with a PID controller that corresponds to the transfer function expressed in (14).

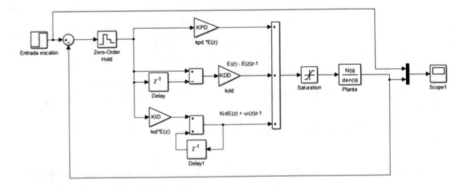

Fig. 4. Block diagram with discrete PID controllers implemented in Matlab.

$$Gc(s) = Kp + \frac{Ki}{s} + Kds = \frac{Kds^2 + Kps + Ki}{s} \tag{14}$$

Obtaining the transfer function in sector loop in the Eq. (15)

$$\frac{Y(s)}{R(s)} = \frac{G_c(s)G_p(s)}{1 + G_c(s)G_p(s)} \tag{15}$$

By operating this expression we come to,

$$\frac{Y(s)}{R(s)} = \frac{(Km)kds^2 + (km)kps + (km)ki}{s^3 + (km)kds^2 + (km)kps + (km)ki} \tag{16}$$

This transfer function obtained is what we must match with the desired model to design the controller.

For our desired model we must choose the value of ρ and Ts for this we make the stabilization time Ts = 4 s and the value of ρ is 0.7 in order to have a small peak value, to find the value of ω o.

$$G(s) = \frac{\omega o^2 * k}{s^2 + (2\rho\omega o^2) + \omega o^2} = \frac{2,689}{s^2 + 2,296s + 2,689} \tag{17}$$

This is the transfer function of our desired model, but the denominator of the transfer function of our total system is of order three and that of our desired model is of order two, so it is necessary to add a pole, (s + p) which must be located at least 5 times further away ($N * \rho * \omega$ o) from the dominant pole of the desired model, it is decided that one pole is 8 times further away;

$$P1 = 9,184 \tag{18}$$

Adding the pole to the desired model,

$$G(s) = \frac{2,689}{s^2 + 2,296s + 2,689} * \frac{9,184}{s + 9,184} \tag{19}$$

$$G(s)\frac{12,82}{S^3 + 15,33S^2 + 38,66S + 12,82} \tag{20}$$

Now it is already possible to match the characteristic polynomial of the system with the desired model in order to find the constants of the controller

$$S^3 + 15,33S^2 + 38,66S + 12,82 = s^3 + (km)kds^2 + (km)kps + (km)ki \tag{21}$$

Equating terms and clearing constants

$$Kd = \frac{15,33}{km}; Kp = \frac{38,66}{km}; Ki = \frac{12,82}{km}; \tag{22}$$

Then proceed to discretize the constants, taking into account that the sampling period can not be less than 20 ms, given the control signal of the actuators.

5   Results

The analysis of the characteristics of the system in closed and open laso was carried out and the following simulation responses were obtained. Once implemented the controller in the system was integrated in the comparison with the results in the simulation, a response with an error in the permanent stable state was obtained due to the existing relationship between the two axes.

It was possible to establish the constants of the controller that allow that the ball does not leave the plate and maintain its position around the established set-point, in addition a sequence of points can be generated that allow to obtain a trajectory. Between these trajectories we have circular, ellipsoid and 8-shaped movements (Fig. 5).

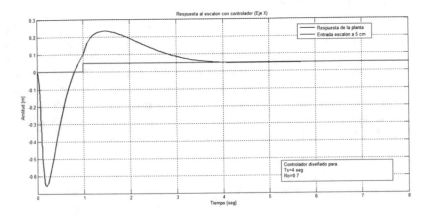

Fig. 5. In response to the scheme in Fig. 4 for the x-axis, the system is tested with a 5 cm step train, where the output signal (blue signal) follows the reference (green signal).

Conclusions

The ball and plate is composed of different elements that were designed by the authors making use of design software and second hand parts with the main objective of creating a low cost platform for academic experimentation of control techniques, a controller has been implemented PID but according to the development of modeling it can be observed that the Ball and Plate system allows an extended research development. Due to this, the costs of the prototype are low compared to systems with the same characteristics that are found in the market, making it accessible to students for academic and didactic purposes.

The inclusion of an application for Android is a complement to the field of control, providing greater versatility among the resources in development, by allowing monitoring through the interface, this being a trend at present as a potential development in home automation and industrial security, monitoring remote, etc. So, applying an Android platform to the ball and plate structure strengthens the character of its development along with the work of control systems.

Acknowledgment. To professors Liliana Fernandez Samaca and Camilo Sanabria Totaitive for guiding us and supporting us in guaranteeing the development of this project.

References

1. Braescu, F.C., Ferariu, L., Gilca, R., Bordianu, V.: Ball on plate balancing system for multi-discipline educational purposes. In: 16th International Conference on System Theory, Control and Computing (ICSTCC), pp. 1–6. IEEE (2012)
2. Farooq, U., Gu, J., Luo, J.: On the interval type-2 fuzzy logic control of ball and plate system. In: International Conference on Robotics and Biomimetics (ROBIO). IEEE (2013). https://doi.org/10.1109/ROBIO.2013.6739804
3. Faruki, P., Bharmal, A., Laxmi, V., Ganmoor, V., Gaur, M.S., Conti, M.: Android security: a survey of issues, malware penetration, and defenses. IEEE Commun. Surv. Tutorials **17**(2), 998–1022 (2015). https://doi.org/10.1109/COMST.2014.2386139
4. Fernandez-Samaca, L., Ramirez, J.M.: An approach to applying project-based learning in engineering courses. In: IEEE ANDESCON. IEEE (2010). https://doi.org/10.1109/ANDESCON.2010.5630007
5. Lee, K.K., Batz, G., Wollherr, D.: Basketball robot: ball-on-plate with pure haptic information. In: 2008 IEEE International Conference on Robotics and Automation. IEEE (2008). https://doi.org/10.1109/ROBOT.2008.4543574
6. Zeeshan, A., Nauman, N., Khan, M.J.: Design, control and implementation of a ball on plate balancing system. In: Proceedings of 2012 9th International Bhurban Conference on Applied Sciences & Technology (IBCAST). IEEE (2012). https://doi.org/10.1109/IBCAST.2012.6177520

Author Index

A. Martínez et al. (Eds.): LACAR 2019, LNNS 112, pp. 267–268, 2020.
https://doi.org/10.1007/978-3-030-40309-6

Printed in the United States
By Bookmasters